A2 PHYSICS
RESOURCE PACK

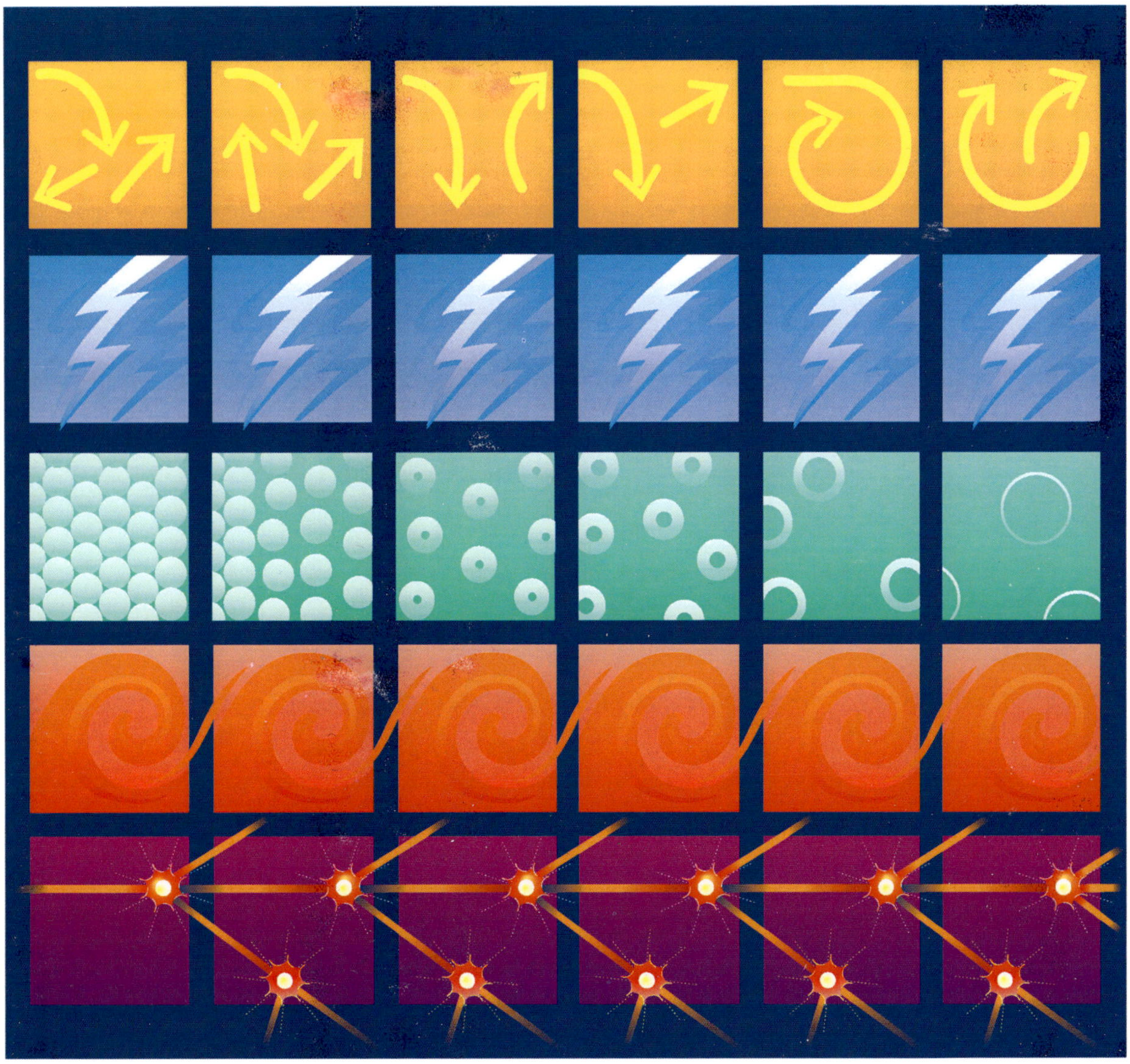

R. HUTCHINGS • M. CRUNDELL • R. KAYE • D. WEBB • E. WEBSTER

A2 PHYSICS RESOURCE PACK

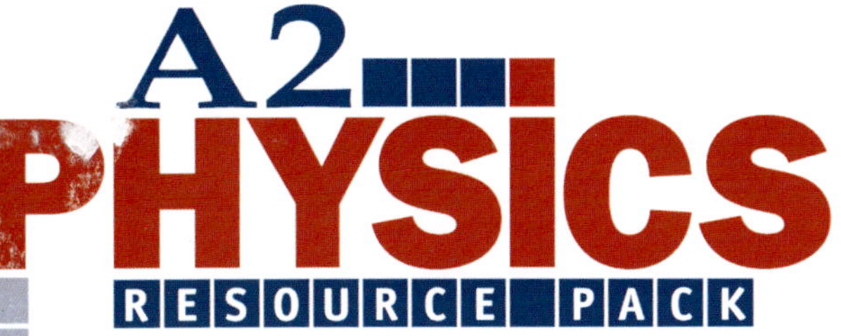

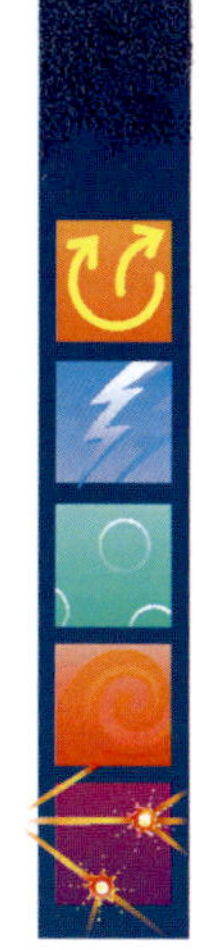

PHILIP ALLAN UPDATES
Market Place
Deddington
Oxfordshire OX15 0SE

Tel: 01869 338652

First published 1999
Revised June 2000

ISBN 0 86003 232 9

Design by Juha Sorsa and Gary Kilpatrick
Printed by Lindsay Ross, Abingdon

Companion title

AS Physics Resource Pack

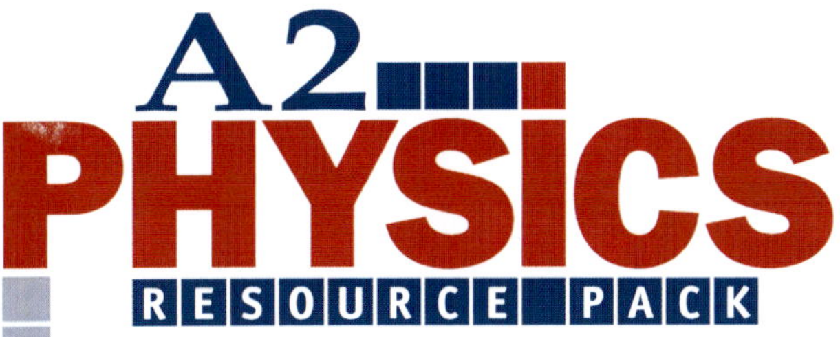

Authors

Robert Hutchings is Chief Examiner for OCR Physics and former Head of Science and Physics at Cheltenham Ladies' College; **Dr Mike Crundell** is Principal Examiner for OCR Physics; **Rosaleen Kaye** teaches at Cheltenham Ladies' College; **David Webb** is an OCR examiner; and **Dr Eric Webster** lectures in the Department of Physics at the University of Central Lancashire, and is an OCR examiner.

Introduction

The aim of this A2 pack and the accompanying AS pack is to provide teachers with prepared lessons for sixthform Physics students. Both are suitable for use in conjunction with any A2 or AS Physics syllabus of the major examination boards.

Each theme in the packs is broken down into many topics and a lesson is provided for each topic. It is assumed that a Physics teacher will be present for most of the time, though it will be necessary for the students to complete many of the exercises in their own time after the lesson.

Timing has not been given for each part of a lesson as this will vary widely from group to group and from school to school, depending on such factors as the background knowledge of the students, their ability and the length of the session available on the school timetable.

Every lesson involves the students doing some work themselves in order to see the significance of the material covered by their teacher. It is important to reach this stage fairly quickly so that applications of the topic can be appreciated. Students can discuss problems amongst themselves and the teacher can help those who encounter difficulties grasping a particular idea.

The packs are not substitutes for textbooks. Much extra detail and answers to specific questions will still need to be found in the standard A-level texts.

The structure of each topic is as follows:

- *Teacher Notes* which provide background information for the topic, *Lesson Pointers* which offer teaching advice (where appropriate) and *Answers to the Student Worksheets*.
- *Student Worksheets* which present series of questions for students to tackle, together with *Basic Facts* where required.

CONTENTS

Section 1 **Mechanics** **1**

Topic 1 Circular motion (1) **2**
Topic 2 Circular motion (2) **6**
Topic 3 Gravitation fields **10**
Topic 4 Newton's law of gravitation **14**
Topic 5 Gravitation potential **19**
Topic 6 Momentum **24**
Topic 7 Impulse **28**
Topic 8 Conservation of linear momentum **31**
Topic 9 Elastic collisions **34**
Topic 10 Inelastic collisions **39**

Section 2 **Electricity** **43**

Topic 1 Fields **44**
Topic 2 Force between point charges **49**
Topic 3 Electric field **52**
Topic 4 Potential gradient **56**
Topic 5 Capacitance **61**
Topic 6 Energy stored in a charged capacitor **65**
Topic 7 Charge and discharge of capacitors **68**
Topic 8 Magnetic effect of an electric current **73**
Topic 9 Electromagnetic induction **78**
Topic 10 Alternating current (a.c.) **85**
Topic 11 The transformer **89**

Section 3 **Heat and Properties of Matter** **94**

Topic 1 Internal energy **95**
Topic 2 Equation of state of an ideal gas **98**
Topic 3 Kinetic theory of gases **101**
Topic 4 Kinetic energy of a molecule of an ideal gas **106**
Topic 5 The ideal gas scale of temperature and absolute zero **110**
Topic 6 First law of thermodynamics **113**

Section 4 **Waves** **118**

Topic 1 Simple harmonic motion **119**
Topic 2 Free and forced vibrations **127**
Topic 3 Wave properties **131**
Topic 4 Polarisation **138**
Topic 5 Diffraction **141**
Topic 6 Double-slit experiment **147**
Topic 7 Stationary waves **153**

Section 5 **Nuclear and Atomic Physics** **158**

Topic 1 Equivalence of mass and energy **159**
Topic 2 Mass defect and binding energy **162**
Topic 3 Fusion **164**
Topic 4 Fission **167**
Topic 5 Wave–particle duality **171**

SECTION
Mechanics

Circular motion (1)

BACKGROUND INFORMATION

Objects that move around in a circle at a constant speed experience a changing velocity since, although the magnitude of the velocity is constant, the direction of the velocity changes constantly. When considering motion in a circle, it is necessary to introduce some method to state how fast the object moves around the circle. The quantity used is the angular velocity, the change in angle per unit time.

THE RADIAN

Angles can be measured in degrees. However, in the case of circular motion, it is found convenient to measure angles using a parameter related to the radius of the circle. This is the unit of angle called the radian. An angle of 1 radian is the angle at the centre of the circle when the arc length s is equal to the radius (see Figure 1).

$$\theta = \frac{\textit{arc length } (s)}{\textit{radius } (r)}$$

usually written in the form $s = r\theta$.

Figure 1

The radian has no units, since it is a length over a length but, so as not to confuse it with angles measured in degrees, the unit rad is often added to angles measured in radians.

If we now consider a circle, the circumference is related to the diameter by π: a circle of radius r has a circumference of $2\pi r$. The angle at the centre of a circle is

$$\theta = \frac{2\pi r}{r} = 2\pi$$

so there are 2π radians at the centre of a circle.

ANGULAR VELOCITY

Angular velocity is the method used to measure the speed of rotation of an object undergoing circular motion.

Angular velocity is the angle turned through, in radians, divided by the time taken, hence units of angular velocity are radians per second (rad s^{-1}).

It is usual to represent the angular velocity by the Greek letter ω, so that, for a particle travelling around a circle with a constant angular velocity

$$\omega = \frac{\textit{angle turned through}}{\textit{time taken}}$$

A particle that makes one revolution of the circle travels through an angle of 2π radians, and the time taken is called the *period of the motion*, T.

$$\omega = \frac{2\pi}{T}$$

Circular motion (1)

If the angular velocity is not constant, we must measure the instantaneous angular velocity

$$\omega = \frac{d\theta}{dt}$$

where $d\theta$ is the angle turned through in a small time dt.

RELATIONSHIP BETWEEN SPEED AND ANGULAR VELOCITY

Consider a particle with a constant speed, moving around the edge of a circle. We cannot say constant velocity since, although the speed is constant, the direction changes all the time and so does the velocity.

The speed of the particle is the distance travelled around the circle divided by the time taken.

$$speed = \frac{distance\ moved\ around\ the\ circumference}{time\ taken} = \frac{ds}{dt} = \frac{r\,d\theta}{dt} = r\omega$$

$$v = r\omega$$

LESSON POINTERS

This section is designed to introduce students to angles measured in radians, and then to use this knowledge to undertake calculations on angular velocity. It can also be used to reinforce the idea that speed is a scalar quantity and velocity a vector quantity.

ANSWERS TO WORKSHEET

1 (a) $\pi/2$ rad
  (b) π rad
  (c) $\pi/3$ rad
  (d) 0.17 rad
2 6.28 s
3 0.122 rad s^{-1}; 51.4 s
4 1.7 rad s^{-1}
5 1.45×10^{-4} rad s^{-1}; 0.105 rad s^{-1}; $+8.1 \times 10^{-8}$ rad s^{-1}; $+5.8 \times 10^{-5}$ rad s^{-1}

Circular motion (1)

BASIC FACTS

The radian

- An angle of 1 radian is the angle at the centre when the arc length s is equal to the radius.

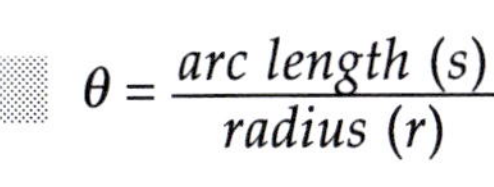

$$\theta = \frac{\textit{arc length } (s)}{\textit{radius } (r)}$$

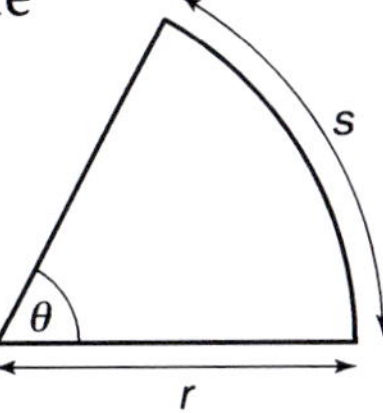

usually written in the form $s = r\theta$.

- The radian has no units, since it is a length over a length but, so as not to confuse it with angles measured in degrees, the unit rad is often added to angles measured in radians.
- The angle at the centre of a circle is

$$\theta = \frac{2\pi r}{r} = 2\pi$$

so there are 2π radians at the centre of a circle.

Angular velocity

- Angular velocity is the angle turned through, in radians, divided by the time taken, hence units of angular velocity are radians per second (rad s^{-1}).

$$\omega = \frac{\textit{angle turned through}}{\textit{time taken}}$$

- A particle that makes one revolution of the circle travels through an angle of 2π, and the time taken is called the period of the motion T.

$$\theta = \frac{2\pi}{T}$$

- If the angular velocity is not constant, we must measure the instantaneous angular velocity

$$\theta = \frac{d\theta}{dt}$$

where $d\theta$ is the angle turned through in a small time dt.

Relationship between speed and angular velocity

- Consider a particle moving around the edge of a circle with a constant speed.

$$\textit{speed} = \frac{\textit{distance moved around the circumference}}{\textit{time taken}} = \frac{ds}{dt} = \frac{r\,d\theta}{dt} = r\omega$$

$$v = r\omega$$

Circular motion (1)

QUESTIONS

1 Convert the following angle at the centre of a circle from degrees to radians: (a) 90°; (b) 180°; (c) 60°; (d) 10°.

2 An object moves around a circle with an angular velocity of 1 rad s^{-1}. What is the period of the motion?

3 An object moving in a circle travels through an angle of 70° in 10 s. Calculate the angular velocity and the period of the motion.

4 A car travels around a corner which has a radius of 20 m; the speed of the car is 34 m s^{-1}. What is the angular velocity of the car?

5 Calculate the angular velocity of the hour hand and the second hand of a clock. The clock is accidentally knocked and is found to gain 2 s every hour. What has been the change in angular velocity of both hands?

Circular motion (2)

BACKGROUND INFORMATION

CENTRIPETAL ACCELERATION

When an object of mass m moves in a circle with constant speed, the velocity is changing because, although the magnitude of the velocity is constant, the direction of the velocity changes.

From Figure 1, we can see that at (1) the velocity is **A** and at (2) the velocity has changed to **C**.

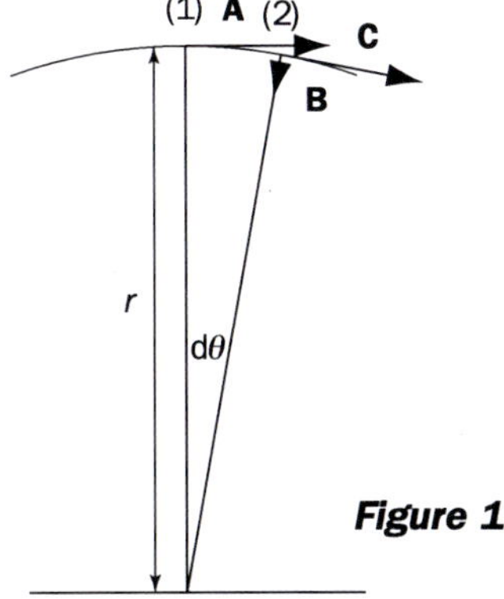

Figure 1

If we now draw a vector diagram of the velocities (Figure 2)

Figure 2

A + *change in velocity* = **C**

Hence, **A** + **B** = **C**, so the change in velocity, **B** = **C** – **A**, which on a vector diagram can be drawn as shown in Figure 3.

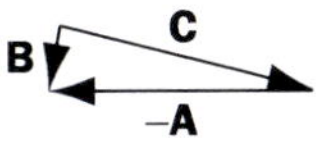

Figure 3

From Figure 3, we can see that the change in velocity **B** is directed towards the centre of the circle.

Since this velocity changes with time there must be an acceleration. The acceleration is in the same direction as the change in velocity, towards the centre of the circle.

This is called the *centripetal acceleration* and, since there is an acceleration towards the centre, there must also be a force on the mass to generate the acceleration: Newton's second law.

It is easy to understand why circular motion requires a force towards the centre of the circle by considering an object on the end of a string, which is made to move in a circle (Figure 4).

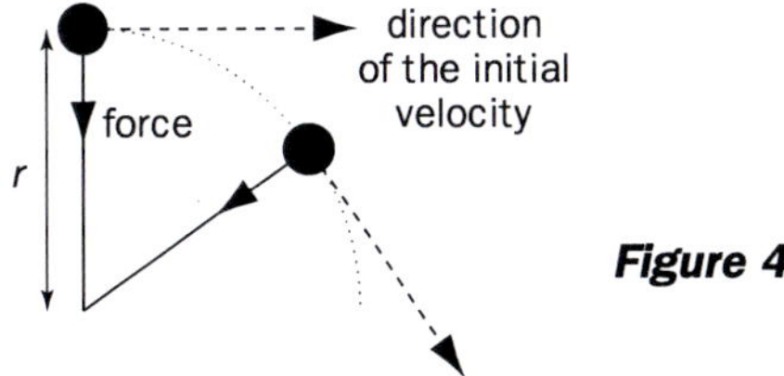

Figure 4

If no force were applied, the object would continue to move along a straight line in the direction of the initial velocity – Newton's first law. To change the direction and so change the velocity, a force must be applied in the direction shown. The same applies at all points around the circle, and so the force is always towards the centre.

Circular motion (2)

CALCULATION OF THE CENTRIPETAL ACCELERATION

In Figure 2, **A** is the initial velocity and **C** is the velocity some small time dt later, when the particle has moved through an angle dθ.

The angle between the two velocities will also be dθ. **B** is the change in velocity required to change velocity **A** to velocity **C** as shown in Figure 5.

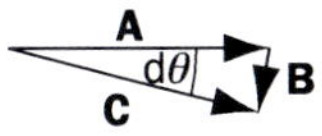

A
dθ
B
C

Figure 5

From Figure 1

$$d\theta = \frac{ds}{r}$$

From Figure 5

$$d\theta = \frac{\mathbf{B}}{\mathbf{A}} = \frac{\textit{change in velocity}}{v} = \frac{dv}{v}$$

Equating these two values of dθ gives

$$\frac{ds}{r} = \frac{dv}{v}$$

The particle moves around the circle with a constant speed v, so that $v = \frac{ds}{dt}$, which is substituted into the above.

$\frac{v dt}{r} = \frac{dv}{r}$ when rearranged gives $\frac{dv}{dt} = \frac{v^2}{r}$, the acceleration towards the centre, the direction of the vector **B**.

Since $v = r\omega$, the centripetal acceleration can be written in three ways:

$$\textit{centripetal acceleration} = \frac{v^2}{r} = r\omega^2 = v\omega$$

CALCULATION OF THE CENTRIPETAL FORCE

A mass m moving in a circle of radius r obeys Newton's second law, so that the *force towards the centre = mass × acceleration towards the centre*:

$$\textit{centripetal force} = \frac{mv^2}{r} = mr\omega^2 = mv\omega$$

CIRCULAR MOTION DUE TO GRAVITY

The Moon orbits the Earth in an approximate circle, and the force towards the centre for this circular motion is provided by the gravitational force between the Earth and the Moon (see Figure 6). The same applies to satellites orbiting the Earth or any other planet.

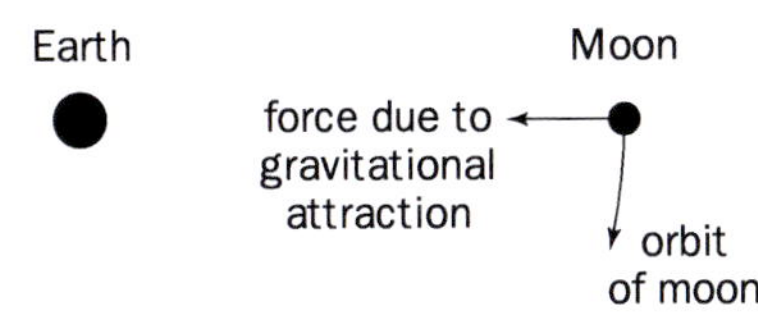

Figure 6

Circular motion (2)

LESSON POINTERS

The concept of an acceleration towards the centre of a circle can be a difficult concept for students to grasp, and it is often easier to introduce the topic by considering the motion of a ball on the end of a string. Starting with Newton's first law, ask students how they would make the mass move in a circle. Using this model, it should be possible to introduce the idea that a force towards the centre is required. It is then a simple step to lead to the second law and show that a force means that an acceleration towards the centre is required.

ANSWERS TO WORKSHEET

1 0.4 rad s^{-1}; 2 m s^{-2}; 0.8 m s^{-2}

2 1.57×10^4 rad s^{-1}; 1.97×10^7 m s^{-2}

3 (a) 1:1
(b) 4:1
(c) 4:1

4 142 m

5 9.9 rad s^{-1}

Circular motion (2)

BASIC FACTS

Centripetal acceleration

- When an object of mass m moves in a circle with constant speed, the velocity is changing because, although the magnitude of the velocity is constant, the direction of the velocity changes.
- Since this velocity changes with time, there must be an acceleration, in the same direction as the change in velocity, towards the centre of the circle.
- This is called the centripetal acceleration.
- *Centripetal acceleration* $= \dfrac{v^2}{r} = r\omega^2 = v\omega$
- Since there is an acceleration towards the centre there must also be a force on the mass to generate the acceleration – Newton's second law.
- *Centripetal force* $= \dfrac{mv^2}{r} = mr\omega^2 = mv\omega$

QUESTIONS

1 An object moves in a circle of radius 5 m with an angular velocity of 0.4 rad s^{-1}. Calculate the speed of the particle tangential to the circle and the centripetal acceleration.

2 A centrifuge rotates at a speed of 2500 rev s^{-1} and has a maximum radius of 8 cm. What are the angular velocity and the maximum centripetal acceleration?

3 A flat disk of radius R rotates at 50 rev s^{-1}. Considering two points on the disk, one at R and one at $R/4$, compare (a) the angular velocities, (b) the tangential speeds, and (c) the centripetal accelerations.

4 A car of mass m travels around a flat curve of radius R, and the friction force between the tyres and the road is 0.45 times the mass of the car. The car is travelling at 25 m s^{-1}. What is the smallest value of R for the car to negotiate the curve without skidding?

5 An object of mass m is rotated in a vertical circle on a string 10 cm long. At the top, the angular velocity is such that the string just remains taut. What is the minimum angular velocity for this to happen?

Gravitation fields

BACKGROUND INFORMATION

FIELDS OF FORCE

The term *field* is used to identify a region of space where a force is experienced. We really should use the term *field of force*, but it is always shortened and just simply called a field. The force generated by the field can have many origins, and perhaps the one with which we are most familiar is the field due to a magnet. From experience we know that, if we bring a nail close to a bar magnet, the nail experiences a force and is attracted to the magnet. We say that the nail is in the field of force, *magnetic field*, of the bar magnet.

Force is a vector quantity and, depending upon the physics of the system, the force will act in a particular direction. In the above example, the force is one that attracts the nail to the magnet. When two magnets are brought together, the force can be either one that attracts the two magnets or one that repels the two magnets.

Hence, whenever an object in a region of space experiences a force, it is said to be in a field of force.

Fields can have many origins; a static charge is surrounded by an electrostatic field and is detected by placing another charge in the field and measuring the force. A moving charge generates an electromagnetic field that can be detected by another moving charge or by a bar magnet. Around an object that has mass, there is a gravitation field, which is detected by placing another mass in the field. In all these cases, the presence of the field is detected by the fact that a force is experienced.

GRAVITATION FIELDS

Gravitation fields, unlike the other fields mentioned, are very weak and require very large masses to generate measurable forces. Surrounding every object that has mass is a gravitation field. However, only when the object has a mass similar in size to that of the Earth is the magnitude of the gravitation field large enough to generate measurable forces when another mass is placed in the field.

The forces generated in gravitation fields are always attractive; the mass placed in the field experiences a force directed toward the mass generating the field.

An object on the surface of the Earth experiences a force towards the centre of the Earth due to the gravitation field of the Earth. The Moon experiences a gravitation force towards the Earth due to the gravitation field of the Earth. The Moon generates its own gravitation field, and it would be equally valid to state that the Earth experiences a force when in the gravitation field of the Moon. In both views, the magnitude of the force would be the same but the direction would be opposite.

A spacecraft travelling from the Earth to the Moon first experiences a force towards the Earth, which decreases as the distance from the Earth increases. At some point during its travel, the gravitation force due to the Earth and the Moon cancel: same magnitude opposite directions.

Gravitation fields

GRAVITATION FIELD STRENGTH

The strength of a gravitation field is measured by placing a unit mass in the field and measuring the magnitude and direction of the force on the object. Hence, by definition, the gravitation field strength is the force on a unit mass placed in the field.

The symbol used for gravitation field strength is g and hence

$$g = \frac{\text{gravitation force}}{\text{mass}} = \frac{F}{m}$$

and g has units of N kg^{-1}.

Other fields are measured in the same way. In the case of an electric field, the electric field strength in magnitude and direction is the force on a unit positive charge placed in the field.

FORCE DIAGRAMS

Around each mass we can draw diagrams of the gravitation field (see Figure 1).

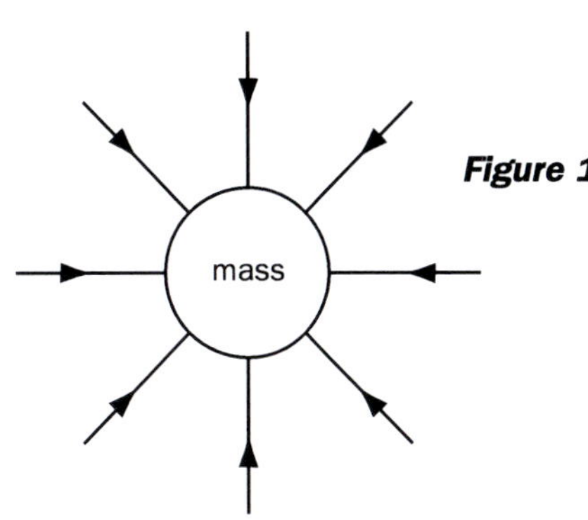

Figure 1

The density of lines represents the magnitude of the field and the arrows the direction. In Figure 1, the gravitation field strength increases the closer we move to the mass.

Gravitation field strength is a vector and so, when gravitation fields are added, the usual rules of vector addition apply.

Gravitation fields

LESSON POINTERS

This section introduces the students to a field of force and students should realise that all fields of force, gravitation, magnetic and electric, are defined by the same formula, the force on some suitable unit object.

At this stage students do not know the way the gravitation field of force varies around a mass. However, they should be able to sketch force graphs and also realise the field is a vector quantity.

ANSWERS TO WORKSHEET

1 Earth Moon Sun

2 Earth Moon

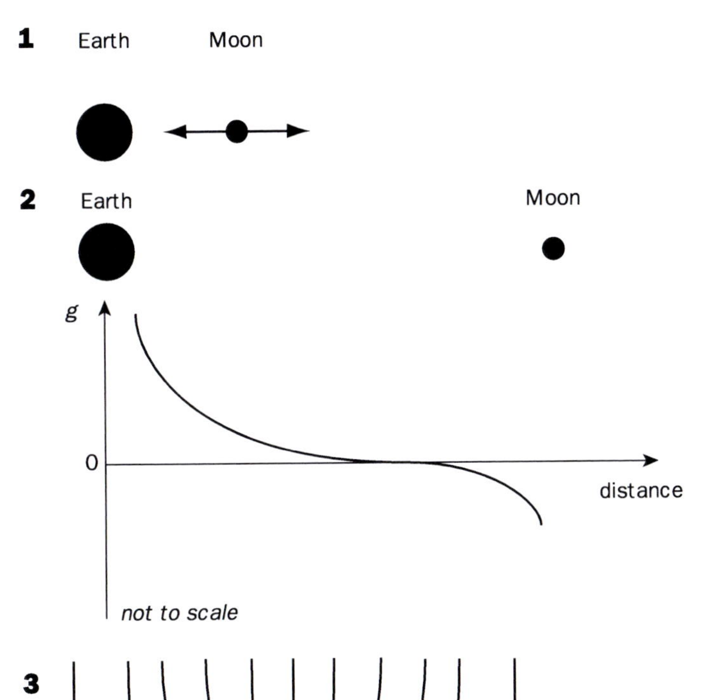

3

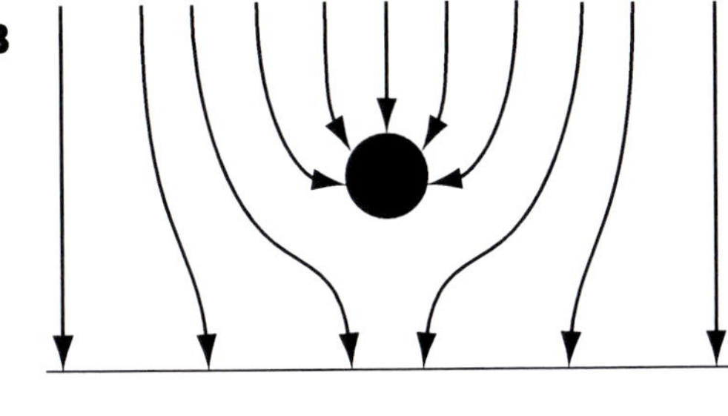

not to scale

4 9.81 N kg^{-1}

5 5.77

Gravitation fields

BASIC FACTS

- Whenever an object in a region of space experiences a force, it is said to be in a field of force.
- Around an object that has mass, there is a gravitation field, which is detected by placing another mass in the field.

Gravitation field strength

- The strength of a gravitation field is measured by placing a unit mass in the field and measuring the magnitude and direction of the force on the object. Hence, by definition, the gravitation field strength is the force on a unit mass placed in the field.
- The symbol used for gravitation field strength is g and hence

$$g = \frac{\text{gravitation force}}{\text{mass}} = \frac{F}{m}$$ and g has units of N kg^{-1}.

Force diagrams

- Around each mass we can draw diagrams of the gravitation field.
- The density of lines represents the magnitude of the field and the arrows the direction. In the diagram on the right, the gravitation field strength increases the closer we move to the mass.
- Gravitation field strength is a vector and so, when gravitation fields are added, the usual rules of vector addition apply.

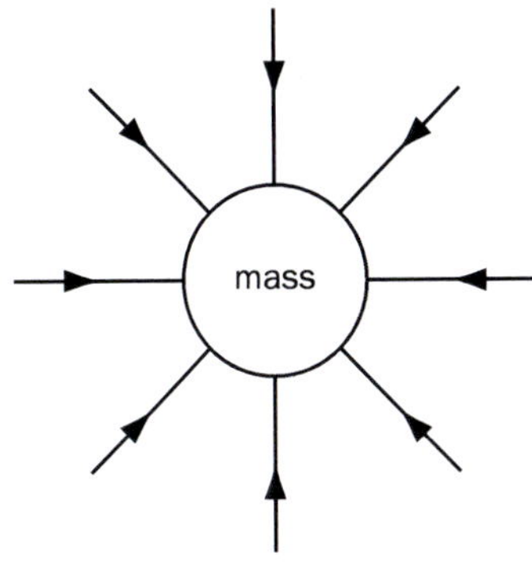

QUESTIONS

1 In a solar eclipse the Sun, Moon and Earth lie in a straight line. Show on a diagram the gravitation forces which are exerted on the Moon at this point and the direction in which they act? A scale diagram is not required.

2 In travelling from the Earth to the Moon, American astronauts experienced gravitation forces. On a suitable graph, sketch the variation in gravitation field strength that they experienced during the trip. A scale diagram is not required.

3 If it were possible to position a huge mass of small size (a neutron ball) just above the Earth's surface, what would be the shape of the gravitation field around the mass? A scale diagram is not required.

4 A mass of 7 kg is placed on a spring balance 1 m above the surface of the Earth and experiences a force of 68.67 N. Calculate the gravitation field strength at that point.

5 The gravitation force on a mass m on the Moon is $1.7m$. Use this information to calculate the ratio of g on the Earth to that on the Moon.

Newton's law of gravitation

BACKGROUND INFORMATION

As covered in the last topic, two objects that have mass experience a force of attraction called gravitation. The magnitude of the force is found to be proportional to:
(1) the masses of the two objects

$$force \propto m_1 \times m_2$$

where m_1 and m_2 are the masses of the two objects.

(2) $\dfrac{1}{(\text{distance between the two objects})^2}$

$$force \propto 1/d^2$$

where d is the distance between the two objects. This is called an *inverse square law*, since the force is proportional to the inverse square of the distance between the two masses. Inverse square laws can also be found in equations relating the forces between charges and magnetic poles.

These two parameters can be combined in one equation:

$$gravitation\ force \propto \frac{m_1 m_2}{d^2}$$

Since m_1 and m_2 are measured in kilograms and d is measured in metres, we must introduce a constant into the equations so that the force can be given in newtons.

$$gravitation\ force = G\frac{m_1 m_2}{d^2}$$ **Newton's law of gravitation**

The constant introduced is G, the gravitation constant, and when experiments are performed it is found to have a value of 6.67×10^{-11} N m^2 kg^{-2}.

The constant, G, is very small, and so gravitation forces are, in general, very small. The gravitation force experienced between two apples placed a distance of 1 m apart is only about 1×10^{-12} N, a very small force indeed, which requires very sophisticated apparatus to detect and measure in the laboratory.

It is only when one of the masses is very large, such as the mass of Earth or the Moon, that the force is large enough to be detected easily.

Newton's law is valid provided the two objects are point masses. Objects with mass cannot be considered point objects. However, Newton was able to show that, provided the object was spherical in shape and the mass uniformly distributed, the force exerted on a similar spherical mass was given by Newton's law, as if all the mass were concentrated at the centre of the sphere and the distances were measured from the centre. This rule applies for points outside the two spheres.

Newton's law of gravitation

GRAVITATION FIELD STRENGTH FROM NEWTON'S LAW

In the last section we saw that the gravitation field strength was defined as

$$g = \frac{gravitation\ force}{mass} = \frac{F}{m}$$

Hence, if a mass m_1 is placed in the gravitation field of a mass m_2, and they are a distance d apart, the gravitation field strength experienced by m_1 can be calculated from Newton's law as follows:

$$g = \frac{gravitation\ force}{mass} = \frac{F}{m_1}$$

$$gravitation\ field\ strength = g = \frac{gravitation\ force}{mass} = \frac{F}{m_1} = G\frac{m_1 m_2}{d^2 m_1} = G\frac{m_2}{d^2}$$

APPLICATION TO A MASS *m* ON THE SURFACE OF THE EARTH

In many calculations we are concerned with masses in the gravitation field of the Earth. In these calculations let M equal the mass of the Earth, m the mass of the object on the surface, and d the distance between the centre of the mass m and the centre of the Earth. Since the Earth is spherical, this is just the radius r of the Earth.

The gravitation field strength is then

$$g = \frac{gravitation\ force}{mass} = \frac{F}{m} = G\frac{mM}{r^2 m} = G\frac{M}{r^2}$$

To a first approximation, the radius of the Earth is constant and, as the mass can be considered as uniformly distributed throughout the sphere, the gravitation field strength on the surface of the Earth is constant.

g AND THE ACCELERATION DUE TO GRAVITY

A mass m on the surface of the Earth falls under the action of gravity and accelerates. The acceleration can be calculated from the gravitation field strength, since Newton's second law of motion states that

$$force = mass \times acceleration$$

A mass m experiences a gravitation force of

$$force = G\frac{mM}{r^2}$$

Using Newton's second law

$$force = G\frac{mM}{r^2} = m \times acceleration\ due\ to\ gravity$$

Hence

Newton's law of gravitation

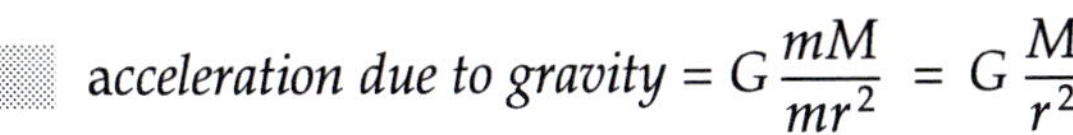

acceleration due to gravity $= G\frac{mM}{mr^2} = G\frac{M}{r^2}$

so

acceleration due to gravity = gravitation field strength

so the two quantities have numerically the same value.

Gravitation field strength = 9.81 N kg^{-1} Acceleration due to gravity = 9.81 m s^{-2}

LESSON POINTERS

This section should not cause much difficulty. It is useful to introduce the concept of inverse square law fields at this point and to mention other inverse square law fields in electrostatics and magnetism.

ANSWERS TO WORKSHEET

1 1.99×10^{20} N

2 2.66×10^{-6} rad s^{-1}; 2.37×10^{6} s or 27.5 days

3 1.6 m s^{-2}

4 4.23×10^{7} m

5 2.0×10^{30} kg

4 Newton's law of gravitation

BASIC FACTS

- As covered in the Topic 3 worksheet, two objects that have mass experience a force of attraction called gravitation. The magnitude of the force is found to be

$$\textit{gravitation force} \propto \frac{m_1 m_2}{d^2}$$

where m_1 and m_2 are the masses of the two objects, and d is the distance between them. Since m_1 and m_2 are measured in kilograms and d is measured in metres, we must introduce a constant into the equations so that the force can be given in Newtons.

$$\textit{gravitation force} = G\frac{m_1 m_2}{d^2}$$ **Newton's law of gravitation**

where G is the gravitation constant and has a value of $6.67 \times 10^{-11}\ \text{N m}^2\ \text{kg}^{-2}$.

- Newton's law is valid provided the two objects are point masses. Objects with mass cannot be considered as point objects. However, Newton was able to show that, provided the object was spherical in shape and the mass uniformly distributed, the force exerted on a similar spherical mass was given by Newton' law, as if all the mass were concentrated at the centre of the sphere and the distances were measured from the centre. This rule applies for points outside the two spheres.

Gravitation field strength from Newton's law

- Gravitation field strength was defined as

$$g = \frac{\textit{gravitation force}}{\textit{mass}} = \frac{F}{m}$$

Hence, if a mass m_1 is placed in the gravitation field of a mass m_2, and they are a distance d apart, the gravitation field strength experienced by m_1 can be calculated from Newton's law as follows:

$$g = \frac{\textit{gravitation force}}{\textit{mass}} = \frac{F}{m_1}$$

$$\textit{gravitation field strength} = g = \frac{\textit{gravitation force}}{\textit{mass}} = \frac{F}{m_1} = G\frac{m_1 m_2}{d^2 m_1} = G\frac{m_2}{d^2}$$

Application to a mass *m* on the surface of the Earth

- The gravitation field strength is given by

$$g = \frac{\textit{gravitation force}}{\textit{mass}} = \frac{F}{m} = G\frac{mM}{r^2 m} = G\frac{M}{r^2}$$

Since the radius of the Earth is constant and the mass can be considered as uniformly distributed, the gravitation field strength on the surface of the Earth is constant.

Newton's law of gravitation

g and the acceleration due to gravity

- A mass m on the surface of the Earth falls under the action of gravity and accelerates. The acceleration can be calculated from the gravitation field strength, since Newton's second law of motion states that *force = mass × acceleration*.
- A mass m experiences a gravitation force of

  $$force = G\frac{mM}{r^2} = m \times acceleration\ due\ to\ gravity$$

  Hence

  $$acceleration\ due\ to\ gravity = G\frac{mM}{mr^2} = G\frac{M}{r^2}$$

  so

  $$acceleration\ due\ to\ gravity = gravitation\ field\ strength$$

  so the two quantities have numerically the same value.
- Gravitation field strength = 9.81 N kg^{-1} Acceleration due to gravity = 9.81 m s^{-2}

QUESTIONS

1 The Moon travels around the Earth at a radius of 3.84×10^8 m, the mass of the Earth is 5.98×10^{24} kg and that of the Moon 7.35×10^{22} kg. Calculate the gravitation force on the Moon due to the Earth.

2 Using the answer to the above question, calculate the angular velocity of the Moon about the Earth and the period of its orbit in days.

3 Calculate the acceleration due to gravity on the Moon. The Moon has radius of 1.74×10^6 m.

4 Communication satellites can be placed in an orbit at a distance from the Earth such that they have an angular velocity the same as the Earth, synchronous orbits. At what distance must the satellite be placed from the Earth for this to be the case?

5 The Earth orbits the Sun at a radius of 1.50×10^{11} m. Calculate the mass of the Sun.

Gravitation potential

BACKGROUND INFORMATION

GRAVITATION POTENTIAL

A mass in a gravitation field experiences a force and, if released, would accelerate and gain kinetic energy. The energy gained is equal to the gravitation potential energy lost as it moves from an initial position in the gravitation field to a final position in the field. When moving in the direction of the field, potential energy is lost, and, when moving against the field, potential energy is gained.

The potential energy lost or gained depends upon the initial and final positions of the mass in the gravitation field. Hence, in a gravitation field there is some property that depends only upon the position of the mass in the field. This is called the *gravitation potential.*

The gravitation potential at a point is defined as the work done in bringing a unit mass from infinity to the point in the field.

The definition implies that the zero of gravitation potential is at infinity, the point at which the gravitation force is also zero. A mass m at infinity, moving towards another mass M, will gain in kinetic energy and so lose potential energy. The gravitation potential around the mass M is always negative.

In most cases, we shall be considering the gravitation potential due to spherical bodies such as the Earth or the planets, in which case the gravitation potential can be easily calculated.

GRAVITATION POTENTIAL DUE TO A SPHERICAL BODY

Consider a mass m moving from point A to point B, a distance δx, in the gravitation field of the spherical object of mass M (Figure 1). If δx is very small, we can consider the force due to the gravitation field to be constant over the distance δx and equal to

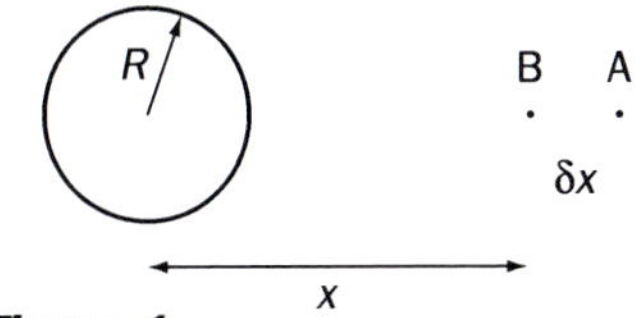

Figure 1

$$force = G\frac{Mm}{x^2}$$

so that

work done on the mass m in moving towards M = *force* × *distance*

$$work\ done\ by\ mass\ m = G\frac{Mm}{x^2}\,\delta x$$

$$work\ done\ on\ a\ unit\ mass = G\frac{Mm}{mx^2}\,\delta x = G\frac{M}{x^2}\,\delta x$$

Hence, the work done on the mass m in moving from infinity to the point x is the integral of the above.

5 Gravitation potential

total work done $= \int_{\infty}^{x} G\frac{M}{x^2}\,\delta x = \left[-G\frac{M}{x}\right]_{\infty}^{x} = \left[-G\frac{M}{x}\right] - \left[-G\frac{M}{\infty}\right]$

total work done = *gravitation potential* $= -G\frac{M}{x}$

Hence, at the surface of the spherical mass M of radius R, the gravitation potential is $-G\frac{M}{R}$. Since gravitation potential is the work done per unit mass, the units are J kg^{-1}.

The gravitation potential energy of a mass m at the surface of the spherical mass is thus $-G\frac{Mm}{R}$ J.

ESCAPE VELOCITIES

In order for a rocket to leave the surface of the Earth and move off into outer space, it must be given an initial kinetic energy. As it moves away from the Earth, this kinetic energy is converted to gravitation potential energy. To escape, the initial kinetic energy must equal the gain in gravitation potential energy in travelling to infinity.

Initial kinetic energy of a mass m = gain in potential energy of a mass m in moving to infinity

$\frac{1}{2}mv^2 = G\frac{Mm}{R}$

Hence, the initial velocity required, the escape velocity, $v = \sqrt{\frac{2GM}{R}}$. In the case of a rocket leaving the Earth, M is the mass of the Earth and R the radius.

GRAVITATION FIELD STRENGTH AND GRAVITATION POTENTIAL

Work done by a unit mass in moving a distance δx against the gravitation field is the increase in gravitation potential δV (Figure 2).

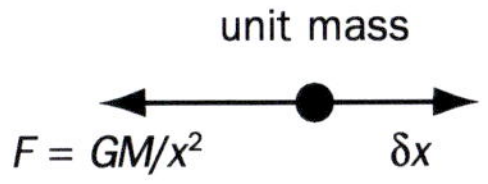

unit mass

$F = GM/x^2$ δx

Figure 2

$\delta V = -G\frac{M}{x^2}\,\delta x$, negative sign because F and δx are in opposite directions.

Hence, $\frac{\delta V}{\delta x} = -G\frac{M}{x^2} = -g$, the gravitation field strength.

The equation is usually written in reverse as $g = -\frac{\delta V}{\delta x}$, gravitation field strength at a point in the field is equal to the negative potential gradient.

Note: When considering electric charges, equivalent equations will be obtained, the only difference being that the forces due to charges can be attractive and repulsive.

Gravitation potential

LESSON POINTERS

The difficult part of this section is the mathematics involved in calculating the gravitation potential. This could be left out and students could just be given the final formula for the gravitation potential. The other difficulty that arises is the concept of choosing the zero of gravitation potential at infinity and that, in moving towards the mass *M*, the gravitation potential will be negative.

ANSWERS TO WORKSHEET

1 -6.25×10^7 J kg^{-1}

2 -6.25×10^8 J; -6.16×10^8 J; 1340 m s^{-1}

3 1.91×10^{11} J kg^{-1}; 6.18×10^5 m s^{-1}

4 9.80 J m^{-1}; 1.62 J m^{-1}

5 1.12×10^4 m s^{-1}, 2.37×10^3 m s^{-1}

5 Gravitation potential

BASIC FACTS

Gravitation potential

- A mass in a gravitation field experiences a force and, if released, would accelerate and gain kinetic energy. The energy gained is equal to the gravitation potential energy lost as it moves from an initial position in the gravitation field to a final position in the field. When moving in the direction of the field, potential energy is lost, and, when moving against the field, potential energy is gained.
- The gravitation potential at a point is defined as the work done in bringing a unit mass from infinity to the point in the field.
- The definition implies that the zero of gravitation potential is at infinity, the point at which the gravitation force is also zero. A mass *m* at infinity moving towards another mass *M* will gain in kinetic energy and so lose potential energy. So the gravitation potential around the mass *M* is always negative.
- Gravitation potential at a point *x* from the centre of a spherical mass *M*:

  total work done in moving a unit mass from infinity to x = *gravitation potential* $= -G\dfrac{M}{x}$

- Hence, at the surface of a spherical mass *M* of radius *R*, the gravitation potential is $-G\dfrac{M}{R}$, the the units are J kg^{-1}.
- For a mass *m* at the surface of a spherical mass *M* of radius *R*, the gravitation potential energy is $-G\dfrac{Mm}{R}$, the unit is J.

Escape velocities

- To escape from the surface of the Earth, the initial kinetic of a mass *m* must equal the gain in gravitation potential energy in travelling to infinity.
- Initial kinetic energy of a mass *m* = gain in potential energy of a mass *m* in moving to infinity

  $$\tfrac{1}{2}mv^2 = G\frac{Mm}{R}$$

- Hence, the initial velocity required, the escape velocity, $v = \sqrt{\dfrac{2GM}{R}}$. In the case of a rocket leaving the Earth, *M* is the mass and *R* is the radius of the Earth.

Gravitation field strength and gravitation potential

- Work done by a unit mass in moving a distance δx against the gravitation field is the increase in gravitation potential δV.
- The gravitation field strength at a point in the field is equal to the negative potential gradient

  $$g = -\frac{\delta V}{\delta x}$$

Gravitation potential

QUESTIONS

1 Calculate the gravitation potential at the surface of the Earth, mass 5.98×10^{24} kg and radius 6.38×10^{6} m.

2 Calculate the gravitation potential energy of a mass of 10 kg on the surface of the Earth and at a height of 100 km above the Earth. If released from this height, with what velocity would the 10 kg mass strike the Earth?

3 Calculate the gravitation potential at the surface of the Sun. The Sun has a mass of 1.99×10^{30} kg and a radius of 6.96×10^{8} m. An asteroid is captured by the Sun after arriving from some distant part of the cosmos. With what velocity would it strike the Sun?

4 Calculate the gravitation potential gradient at the surface of the Earth, mass 5.98×10^{24} kg and radius 6.38×10^{6} m, and the Moon, mass 7.35×10^{22} kg and radius 1.74×10^{6} m.

5 Calculate the escape velocity of a rocket from the surface of the Earth and compare it with that required for leaving the surface of the Moon.

Momentum

BACKGROUND INFORMATION

The momentum of a body of mass m is defined as the product of the mass and the velocity, v, of the body. Since velocity is a vector quantity, momentum is also a vector quantity, and we must always state the direction of the momentum as well as the magnitude.

$$momentum = p = mv$$

Newton's second law of motion uses momentum to describe the effect of forces on bodies.

REVISION OF FORCE NOTES

Newton's second law

The rate of change of momentum of a body is proportional to the total force acting on it and takes place in the direction of the force.

Consider an object of mass m acted upon by a force F so that its velocity increases in magnitude from u to v in a time of t seconds.

$$change\ of\ momentum/time \propto force$$

$$\frac{mv - mu}{t} \propto \mathrm{F}$$

$$\frac{m(v - u)}{t} \propto \mathrm{F}$$

To remove the proportionality sign, we introduce a constant.

$$\mathrm{C}\,\frac{m(v - u)}{t} = \mathrm{F}$$

It is usual to write this equation in reverse order as

$$\mathrm{F} = \mathrm{C}\,\frac{m(v - u)}{t}$$

If the force is measured in newtons, the constant is assigned a value of 1.

$$\mathrm{F} = \frac{m(v - u)}{t}$$

so that the force in newtons equals the rate of change of momentum.

MOMENTUM OF A BODY MOVING IN A CIRCLE

A body of mass m moving in a circle with a constant speed v also experiences a change in momentum; in this case the magnitude of the momentum is constant but the direction changes as the particle moves around the circle. The rate of change of momentum gives the force towards the centre required for circular motion. In this case, calculations are best performed using vector diagrams.

Momentum

Consider the diagram used in the notes on circular motion. The length of the lines drawn gives the magnitude of the momentum, and they are the same length at the two points on the circle, as the magnitude is the same. The momentum, however, is different, because they point in different directions (see Figure 1).

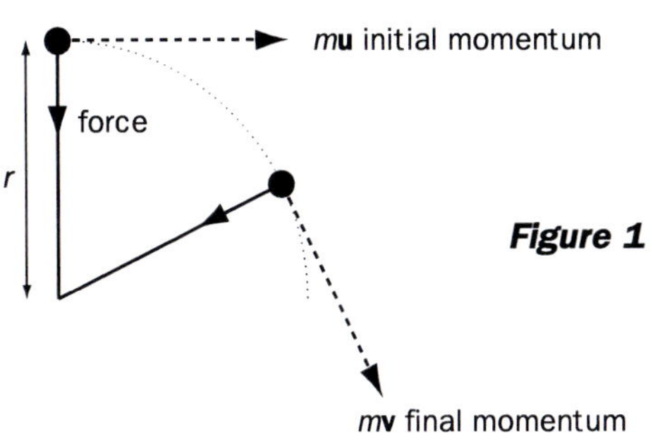

Figure 1

The change in momentum can be obtained from a vector diagram (Figure 2).

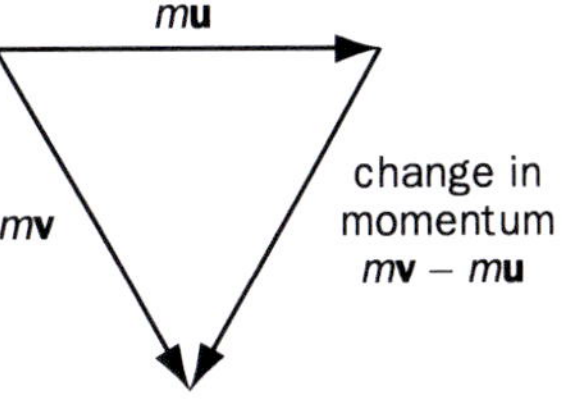

Figure 2

In the vector diagram the change in momentum is given by the following equation

final momentum = initial momentum + change in momentum

$m\mathbf{v} = m\mathbf{u} + (m\mathbf{v} - m\mathbf{u})$

The direction of the change in momentum, $m\mathbf{v} - m\mathbf{u}$, is towards the centre of the circle, which is the direction in which the force acts, thus agreeing with Newton's second law as applied to circular motion.

EXAMPLES OF OTHER CHANGES IN MOMENTUM

There are many examples in physics where a change in momentum provides a force. Water leaving a hose-pipe suffers a change in momentum when the water, which is moving relatively slowly in the pipe, speeds up rapidly as it exits through the nozzle.

A jet engine works in the same way, with gas being drawn in at the front at a slow velocity to be heated and ejected at a very fast velocity at the exhaust.

Momentum

LESSON POINTERS

This section is a revision of the force notes, except that the use of Newton's second law generates an equivalent version of the $F = ma$ equation in terms of rate of change of momentum. Students need to appreciate that this is just $F = ma$ in a different form, which is easier to apply to certain physical situations. The section also provides a revision of vectors by looking at drawing vector diagrams of momentum for motion of an object in a circle.

Practical applications of this form of Newton's law are introduced.

ANSWERS TO WORKSHEET

1 35 000 kg m s^{-1}; 47 600 kg m s^{-1}; 12 600 kg m s^{-1}

2 564 kg m s^{-1}; 63°E of S

3 1.8 kg m s^{-1}, upwards

4 485 N

5 108 N

Momentum

BASIC FACTS

- The momentum of a body of mass m is defined as the product of the mass and the velocity v of the body. Since velocity is a vector quantity, momentum is also a vector quantity, and we must always state the direction of the momentum as well as the magnitude.
- *momentum* $= p = mv$
- Newton's second law of motion uses momentum to describe the effect of forces on bodies.
- Newton's second law

  The rate of change of momentum of a body is proportional to the total force acting on it and takes place in the direction of the force.
- $F = \dfrac{m(v - u)}{t}$
- Force in newtons equals rate of change of momentum.

QUESTIONS

1 A car of mass 1400 kg is moving with a velocity of 25 m s^{-1}. What is its momentum? The car then accelerates to a speed of 34 m s^{-1}. What is its new momentum and what is the change in momentum?

2 A boat of mass 63 kg sails west with a speed of 8 m s^{-1}. It then changes direction and sails south with a speed of 4 m s^{-1}. Using a scale vector diagram, calculate the magnitude and direction of the change in momentum.

3 A ball of mass 200 g falls to the ground and just before striking the ground its speed is 5 m s^{-1}. It rebounds with a velocity of 4 m s^{-1}. What is the magnitude and direction of the change in momentum?

4 The car in question 1 accelerates from 25 m s^{-1} to 34 m s^{-1} in a time of 26 s. What is the average force applied?

5 A water hose ejects water in a fine jet at a rate of 250 litres per minute and at a speed of 26 m s^{-1}. The density of water is 1000 kg m^{-3}. What force is required to hold the hose steady?

7 Impulse

BACKGROUND INFORMATION

In the previous section we looked at Newton's second law in the form

$$F = \frac{m(v - u)}{t}$$

Rearranging this equation gives

$$Ft = m(v - u)$$

The term Ft is called the *impulse*.

So, when a force F is applied to a body for a time t, it is equal to the change in momentum and equals the impulse.

FORCE–TIME GRAPHS

When forces are applied to objects, it is often convenient to describe the system using a force–time graph (Figure 1).

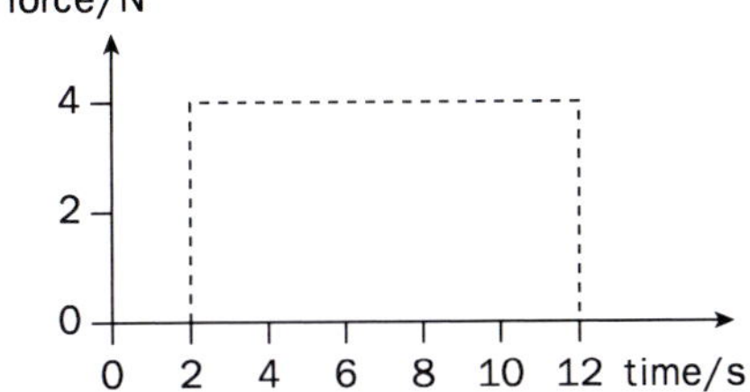

Figure 1

Consider an object subject to a constant force of 4 N applied for 10 s. We can use this information to draw a force–time graph.

The change in momentum is then the force × the time for which the force is applied, and is the area under the graph.

$$\textit{change in momentum} = 4 \times 10 = 40 \text{ kg m s}^{-1}$$

FORCE–TIME GRAPHS FOR NON-CONSTANT FORCES

The above is a rather trivial calculation, since the force is constant, but there are many examples in physics where the force changes with time. One such simple example is when a stationary ball is hit with a bat, where the force–time graph would probably have the shape shown in Figure 2.

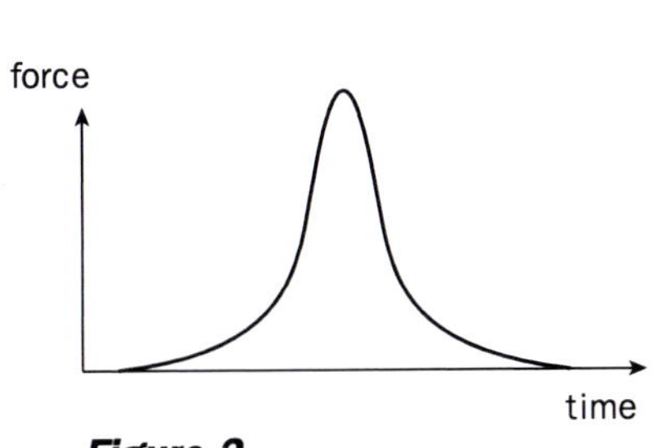

Figure 2

It can be seen that the force rises to a maximum value when the ball is compressed by the greatest amount, and the falls to zero as the ball moves away from the bat. The same rule applies as for a constant force, and the change in momentum is given by the area under the curve.

Impulse

LESSON POINTERS

This section is a simple extension of the section on momentum and should not cause many difficulties.

ANSWERS TO WORKSHEET

1 1.89 N s; 945 N

2 36.4 m s^{-1}

3 0.050 N

4 6.3×10^{-4} s; 0.475 N s; 754 N

5 14 m s^{-1}; 1120 N s; 747 N

Impulse

BASIC FACTS

- If we take the equation $F = \frac{m(v - u)}{t}$, rearranging gives $Ft = m(v - u)$.
- The term Ft is called the impulse.

Force-time graphs

- The change in momentum is Ft, the area under the graph.

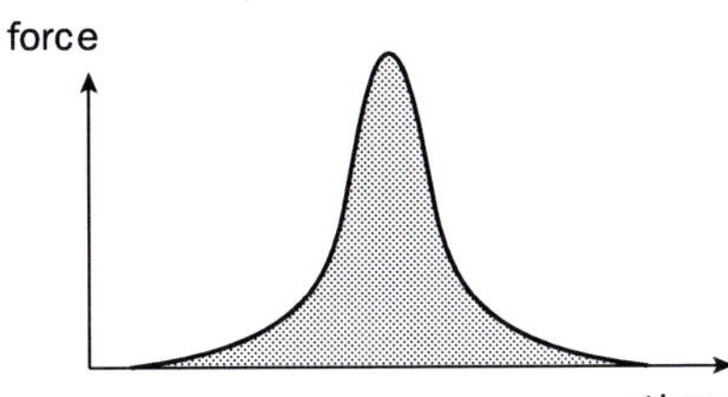

Force-time graphs for non-constant forces

- Force–time graph for hitting a stationary ball with a bat.
- The change in momentum is given by the area under the curve.

QUESTIONS

1 When a golf ball of mass 45 g is hit by a golf club, it leaves the club head with a speed of $42\,\text{m}\,\text{s}^{-1}$. The club head is in contact with the ball for 2.0×10^{-3} s. Calculate the impulse and the average force exerted by the club on the ball.

2 The force–time graph of a racket hitting a tennis ball can be drawn simply as shown.

If the ball is initially stationary prior to being hit by the racquet, with what speed does the ball leave the racket? A tennis ball has a mass of 55 g.

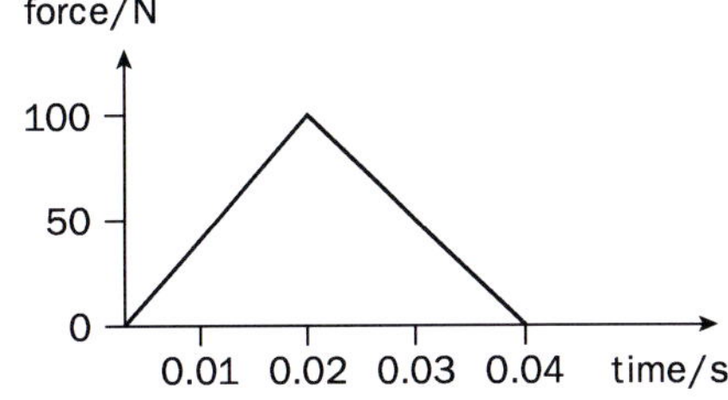

3 Raindrops strike the ground with a velocity of 10 m s^{-1}. In a typical rain shower there are 50 g of water in a cubic metre of air. What is the force which the raindrops exert on an area of ground of 100 cm^2?

4 A 5 g bullet travelling at 95 m s^{-1} strikes a wooden target and comes to rest with uniform deceleration in a distance of 3 cm. Calculate the time taken for the bullet to stop, the impulse and the average retarding force.

5 A high diver of mass 80 kg dives into a pool from the 10 m board and comes to rest in the pool in a time of 1.5 s. Assume the initial vertical velocity is zero. With what speed does the diver enter the water, and what is the impulse and average force provided by the water in the pool?

Conservation of linear momentum

BACKGROUND INFORMATION

When two objects are in collision and no external forces act, from Newton's third law, the force on one object must be equal and opposite to the force on the other. If they are in collision for a time t, we can calculate the impulse during the collision as $force \times t$. The force F may be a constant force or, as in most practical examples, the force will vary with time. With two objects, the force on each object will have the same magnitude but the force will act in opposite directions.

For the first object: $F_1t = m_1v_1 - m_1u_1$.

For the second object: $F_2t = m_2v_2 - m_2u_2$.

Since the forces on each object are the same but act in opposite directions, $F_1t = -F_2t$.

Substituting from the above gives $m_1v_1 - m_1u_1 = -(m_2v_2 - m_2u_2)$.

Rearranging gives $m_1u_1 + m_2u_2 = m_1v_1 + m_2v_2$.

The total momentum of the objects after the collision is equal to the total momentum of the objects before the collision.

Since momentum is a vector quantity, a change in momentum may involve a change in direction as well as a change in magnitude.

This is known as the *principle of conservation of linear momentum.* It applies to all collisions in physics, from collisions between the smallest atomic particles in particle accelerators to the collision of very large objects, such as the collision between two ships at sea.

- *The total linear momentum of a system of interacting bodies, when no external forces act, remains constant.*

The simplest conservation of linear momentum calculations are performed when the direction of motion of the two objects before and after the collision are along the same straight line. In this case, the calculation just involves deciding in which direction the momentum is going to be considered to be positive. Momentum in the reverse direction is then negative.

Example

When a radium atom, which is initially at rest, emits an alpha particle and changes to a radon atom, the law of conservation of linear momentum applies (see Figure 1).

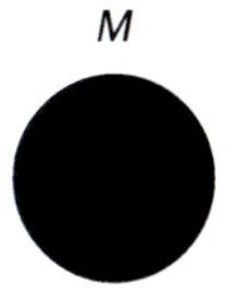

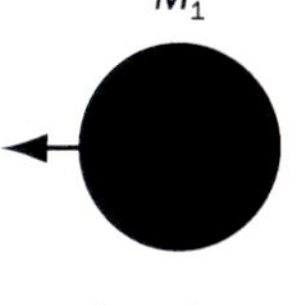

Figure 1

momentum before = momentum after

$0 = m_2v_2 - M_1v_1$

M_1 is the mass of the radon atom, and m_2 is the mass of the alpha particle.

Conservation of linear momentum

$$v_2 = \frac{M_1}{m_2} v_1$$

Since M_1 has a mass of 266 u and the alpha particle a mass of 4 u, $v_2 = 66.5v_1$. The alpha particle moves off very rapidly compared with the radon atom. The same calculation can be performed for a bullet leaving a gun. Initially, both are stationary and, when fired, the bullet moves in one direction with a very high velocity and the gun, since it has a larger mass, moves back with a small velocity – the recoil.

When the collision between the particles involves a change in direction, since momentum is a vector quantity, the law must be applied by considering the conservation of momentum along suitable directions. The directions chosen are usually dictated by the physics of the problem.

Figure 2

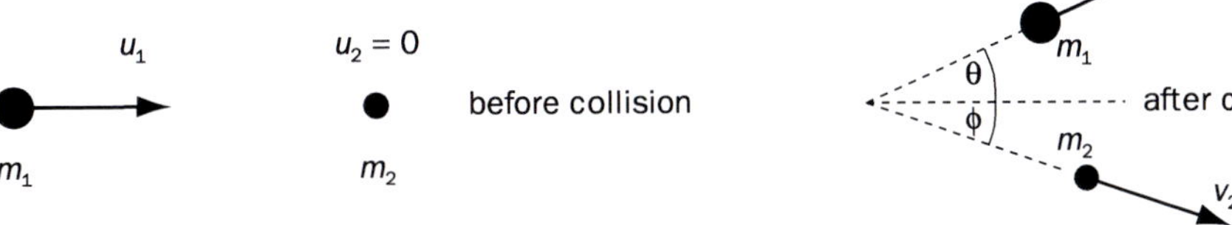

In Figure 2, where an object m_1 strikes a stationary object m_2 and they both move off at an angle, the directions chosen to perform calculations using the conservation of linear momentum are the horizontal and vertical directions.

The equations of conservation of linear momentum are then:
in the horizontal direction

$$m_1u_1 = m_1v_1 \cos\theta + m_2v_2 \cos\phi$$

and in the vertical direction

$$0 = m_1v_1 \sin\theta\tfrac{1}{2} - m_2v_2 \sin\phi$$

Momentum up is taken as positive, and momentum down as negative.

LESSON POINTERS

Conservation of linear momentum calculations need only be considered in one direction. The notes on two-dimensional collisions are included to reinforce the idea that momentum is a vector quantity.

ANSWERS TO WORKSHEET

1 1.1 m s^{-1}

2 7 m s^{-1}

3 14.8 m s^{-1}; 3.3×10^5 J

4 0.75 m s^{-1} in the opposite direction to the woman

5 $1.38 \times 10^{-2} \text{ m s}^{-1}$ in the opposite direction to the bullet

Conservation of linear momentum

BASIC FACTS

- When two objects are in collision and no external forces act, the collision is governed by the principle of the conservation of linear momentum.
- $m_1u_1 + m_2u_2 = m_1v_1 + m_2v_2$
- The total momentum of the objects after the collision is equal to the total momentum of the objects before the collision.
- Since momentum is a vector quantity, a change in momentum may involve a change in direction as well as a change in magnitude.
- This is known as the principle of the conservation of linear momentum. It applies to all collisions in physics, from collisions between the smallest atomic particles in particle accelerators to the collision of very large objects, such as the collision between two ships at sea.
- *The total linear momentum of a system of interacting bodies, when no external forces act, remains constant.*
- The simplest conservation of momentum calculations are performed when the directions of motion of the two objects before and after the collision are along the same straight line. In this case, the calculation just involves deciding in which direction the momentum is going to be considered to be positive. Momentum in the reverse direction is then negative.
- When the collision between the particles involves a change in direction, since momentum is a vector quantity, the law must be applied by considering the conservation of momentum along suitable directions. The directions chosen are dictated by the physics of the problem.

QUESTIONS

1 In an old wooden warship, the cannons were mounted on wheels and had a mass of 1500 kg. The cannons fired cannon balls with a mass of 4.4 kg, which left the cannon with a horizontal velocity of 360 m s^{-1}. What was the recoil velocity of the cannon when it was fired?

2 A ball of mass 2 kg and velocity 7 m s^{-1} makes a head-on collision with a second ball of mass 4 kg moving in the same direction with a velocity of 4 m s^{-1}. After the collision, the velocity of the 2 kg ball is reduced to 1 m s^{-1}. With what velocity does the 4 kg ball move off?

3 A car of mass 1200 kg, travelling at 32 m s^{-1}, collides with a stationary car of mass 1400 kg. After the collision they are locked together and so move off with the same velocity. Calculate the velocity and the change in kinetic energy of the system.

4 A woman of mass 50 kg stands in a boat of mass 80 kg. Initially both are stationary. If the woman walks to one end of the boat with a velocity of 1.2 m s^{-1}, with what velocity does the boat move, and in which direction does it move?

5 A gunman fires a bullet at a moving block of wood. The bullet leaves the gun with a velocity of 520 m s^{-1} and has a mass of 15 g. The block of wood has a mass of 2.5 kg and is moving towards the gunman with a velocity of 6 m s^{-1}. After striking the block of wood, the bullet comes to rest in the block of wood. What is the new velocity of the block of wood?

9 Elastic collisions

BACKGROUND INFORMATION

Collisions between bodies can be classified into one of two types:

1 *elastic collisions*, in which both linear momentum and kinetic energy are conserved;

2 *inelastic collisions*, in which linear momentum is conserved but kinetic energy is not conserved.

All practical collisions fall into the second category and, in general, some kinetic energy is lost in the collision, having been converted into some other form of energy such as sound or heat.

Collisions on an atomic scale can on occasions approximate to elastic collisions.

HEAD-ON ELASTIC COLLISION BETWEEN MOVING AND STATIONARY OBJECTS

Consider an object of mass m_1 colliding with an object of mass m_2. The mass m_1 has an initial velocity u_1 and the mass m_2 is stationary. Figures 1 and 2 show the objects before and after the collision.

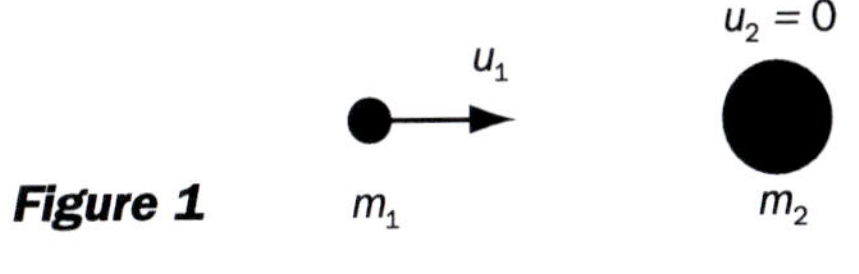

Figure 1

If this is an elastic collision, we can apply the laws of conservation of linear momentum and kinetic energy.

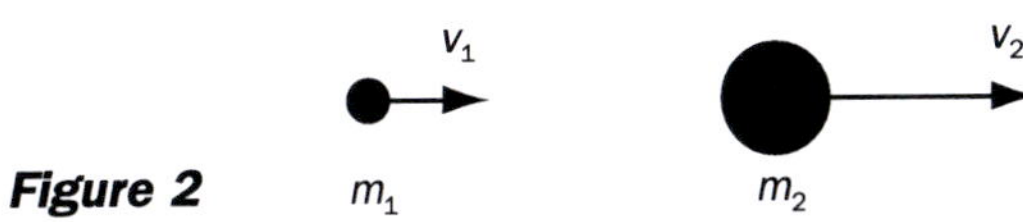

Figure 2

Conservation of linear momentum

In applying this equation, we will consider momentum to the right to be positive.

$$m_1u_1 + 0 = m_1v_1 + m_2v_2 \qquad (1)$$

Conservation of kinetic energy

$$\tfrac{1}{2}m_1u_1^2 + 0 = \tfrac{1}{2}m_1v_1^2 + \tfrac{1}{2}m_2v_2^2 \qquad (2)$$

We have two equations and two unknowns, v_1 and v_2, and can solve these two equations to find the two unknown velocities as follows. To begin with, eliminate v_2, using the two equations.

From (1)

$$m_1u_1 - m_1v_1 = m_2v_2 \qquad (3)$$

From (2)

$$m_1u_1^2 - m_1v_1^2 = m_2v_2^2 \qquad (4)$$

Dividing (4) by (3) gives

9 Elastic collisions

$$\frac{u_1^2 - v_1^2}{u_1 - v_1} = v_2 \qquad (5)$$

but $(u_1^2 - v_1^2) = (u_1 - v_1)(u_1 + v_1)$

$$\frac{(u_1 - v_1)(u_1 + v_1)}{(u_1 - v_1)} = v_2$$

$$u_1 + v_1 = v_2 \qquad u_1 = v_2 - v_1 \qquad (6)$$

This equation shows that, in an elastic head-on collision, the velocity with which two bodies approach each other is equal to the velocity with which they separate from each other.

It is now possible to calculate v_1 by eliminating v_2 from the equations. From (3)

$$m_1u_1 - m_1v_1 = m_2v_2$$

and from (6)

$$u_1 = v_2 - v_1$$

Then

$$m_1u_1 - m_1v_1 = m_2(u_1 + v_1)$$

$$(m_1 - m_2)u_1 = (m_1 + m_2)v_1$$

so that

$$v_1 = \left(\frac{m_1 - m_2}{m_1 + m_2}\right)u_1$$

From this equation we can see that

- if $m_2 > m_1$, the velocity of m_1 is negative, so the object bounces back off the mass m_2;
- if $m_2 \gg m_1$, after the collision $v_1 = -u_1$, and $v_2 = 0$;
- if $m_2 = m_1$, the mass m_1 is stationary after the collision and the mass m_2 moves off with the same velocity as the initial velocity of m_1;
- if $m_2 < m_1$, both particles move off in the same direction as m_1, but m_1 has a reduced velocity.

LESSON POINTERS

This section is an extension of the previous section on the conservation of linear momentum, with just an additional conservation rule to be considered. Again students need to be aware about the vector nature of momentum.

Elastic collisions

ANSWERS TO WORKSHEET

1 $u/2$; $-u/2$

2 -0.040 m s^{-1}; 0.16 m s^{-1}

3 0.029 m s^{-1}; 0.229 m s^{-1}; 3.1 mJ

4 $m_1 = 13.9m_2$

5 With hydrogen $v = 0$, with carbon $v = 846$ u, so hydrogen is more effective in slowing down neutrons. Fraction of energy of neutron lost is 1.0 for hydrogen and 0.284 for carbon.

9 Elastic collisions

BASIC FACTS

Elastic collisions

- Collisions between bodies can be classified into one of two types:
  1 *elastic collisions*, in which both linear momentum and kinetic energy are conserved.
  2 *inelastic collisions*, in which linear momentum is conserved but kinetic energy is not conserved.
- All practical collisions fall into the second category and, in general, some kinetic energy is lost in the collision, having been converted into some other form of energy such as sound or heat.
- Collisions on an atomic scale can on occasions approximate to elastic collisions.

Elastic collision between a moving and stationary object

- Before the collision

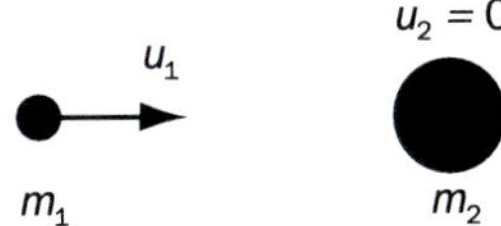

- After the collision

- If this is an elastic collision, we can apply the laws of conservation of linear momentum and kinetic energy.
- *Conservation of linear momentum*

$$m_1u_1 + 0 = m_1v_1 + m_2v_2 \quad (1)$$

- Conservation of kinetic energy

$$\tfrac{1}{2}m_1u_1^2 + 0 = \tfrac{1}{2}m_1v_1^2 + \tfrac{1}{2}m_2v_2^2 \quad (2)$$

Solving these equations for v_1 and v_2 gives

$$u_1 = v_2 - v_1$$

This equation shows that the velocity with which two bodies approach each other in a head-on collision is equal to the velocity with which they separate from each other.

$$v_1 = \left(\frac{m_1 - m_2}{m_1 + m_2}\right) u_1$$

From this equation we can see that:

if $m_2 > m_1$, the velocity of m_1 is negative, so the object bounces back off the mass m_2;

Elastic collisions

if $m_2 \gg m_1$, after the collision $v_1 = -u_1$, and $v_2 = 0$;

if $m_2 = m_1$, the mass m_1 is stationary after the collision and the mass m_2 moves off with the same velocity as the initial velocity of m_1;

if $m_2 < m_1$, both particles move off in the same direction as m_1, but m_1 has a reduced velocity.

QUESTIONS

1 A particle of mass m and velocity u collides elastically and head-on with a second stationary particle of mass $3m$. What will be the velocity of the two particles after the collision? Show that they separate from each other with a velocity u.

2 On an air track in the laboratory, a moving slider of mass 80 g and initial velocity 0.20 m s^{-1} is in elastic collision with a stationary slider of mass 120 g. Calculate the velocity of both sliders after the collision.

3 If in the above question the mass of the moving slider is doubled, what are the new velocities after the collision? How much kinetic energy has been lost by the incident slider?

4 In a head-on elastic collision between two masses, one of which is stationary, an experimenter would like the incident mass to lose one quarter of its kinetic energy in the collision. What must be the ratio of the masses of the two objects for this to be the case?

5 In a nuclear reactor, collisions take place between neutrons and a material called a moderator. Two types of moderator are used in reactors: water or graphite (carbon). If the collisions can be considered as elastic, what fraction of the energy of the neutron is lost in a head-on collision with each atom? Hence, which moderator slows the neutron down fastest? Take the mass of the neutron and the hydrogen nuclei to be 1 u and the carbon nucleus 12 u.

10 Inelastic collisions

BACKGROUND INFORMATION

In inelastic collisions, the kinetic energy in the collision is not conserved. However, linear momentum is conserved in all collisions.

In an inelastic collision, some of the kinetic energy of the colliding objects is lost to other forms of energy. Inelastic collisions are the collisions experienced in everyday life. A ball falling to the ground and bouncing experiences an inelastic collision with the ground, so that, on leaving the ground, some of its kinetic energy has been lost as heat to the ground and as sound, and so the ball leaves with a lower velocity. As time goes on, with each bounce it loses some energy, until all the energy has been dissipated and the ball stops bouncing. If collisions in the real world were perfectly elastic, the ball would continue to bounce for ever and we would have perpetual motion.

Hence, all the collisions we experience between objects are inelastic collisions and some of the kinetic energy is lost in the collision.

ENERGY LOST IN AN INELASTIC COLLISION

Using the conservation of momentum and energy equations, it is possible to calculate the energy lost to other forms such as heat, sound etc.

Consider the head on collision used previously. Figures 1 and 2 show the objects before and after the collision.

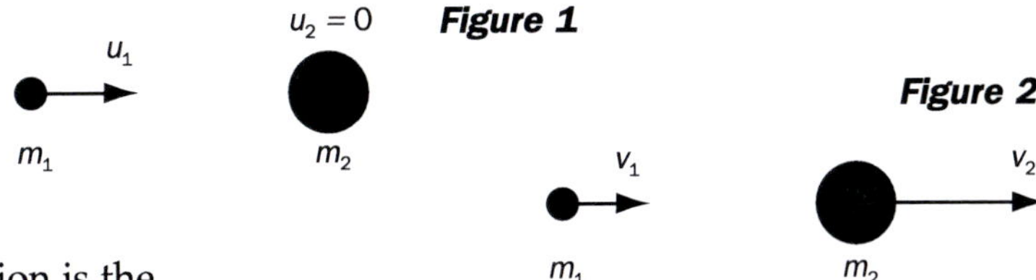

Figure 1

Figure 2

The conservation of momentum equation is the same as before:

$$m_1u_1 + 0 = m_1v_1 + m_2v_2 \qquad (1)$$

The conservation of kinetic energy equation must now be modified as follows:

$$\tfrac{1}{2}m_1u_1^2 + 0 = \tfrac{1}{2}m_1v_1^2 + \tfrac{1}{2}m_2v_2^2 + energy\ lost \qquad (2)$$

In this system there are now three unknowns: v_1, v_2 and the energy lost. In order to solve the problem, we must know at least one of the three unknowns.

10 Inelastic collisions

In many collisions, however, the two colliding objects stick together, in which case their final velocity is the same for m_1 and m_2 and the common speed can be found by a simple application of the conservation of momentum. Such collisions are often called completely inelastic collisions.

Consider the collision between two objects shown in Figure 3.

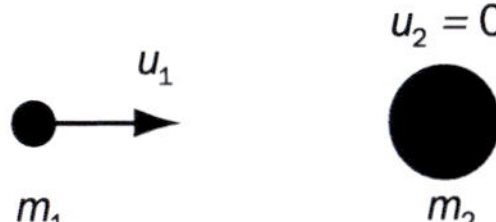

Figure 3

If, after the collision, they stick together and move with a common velocity (Figure 4), its value is then given by the equation $m_1u_1 + 0 = (m_1 + m_2)v_2$ so that

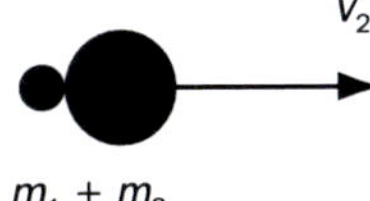

Figure 4

$$v_2 = \frac{(m_1)}{(m_1 + m_2)} u_1.$$

Once v_2 has been calculated, it is possible to calculate the energy lost in the collision, which is given by the equation

$$\tfrac{1}{2}(m_1 + m_2)v_2^2 - \tfrac{1}{2}m_1u_1^2.$$

If both objects are moving before the collision (Figure 5), the conservation of momentum equation contains an extra component. Remember that momentum is a vector quantity and the correct signs must be used.

Figure 5

Conservation of momentum in this case is $m_1u_1 - m_2u_2 = (m_1 + m_2)v_2$. v_2 would then be positive or negative, depending upon the masses and velocities of the colliding objects.

LESSON POINTERS

Collisions in this section are again just in one dimension, and all the calculations involve the two objects joining together after the collision.

ANSWERS TO WORKSHEET

1 1 m s^{-1}; 12 J

2 –5 m s^{-1}; 108 J

3 17.9 m s^{-1}; 2.75×10^5 J

4 –2.3 km h^{-1}; 1.66×10^4 J

5 615 m s^{-1}

Inelastic collisions

BASIC FACTS

- In inelastic collisions, the kinetic energy in the collision is not conserved. However, linear momentum is conserved in all collisions.
- All the collisions we experience between objects are inelastic collisions and so some of the kinetic energy is lost in the collision.

Energy lost in an inelastic collision

- Before the collision
- After the collision

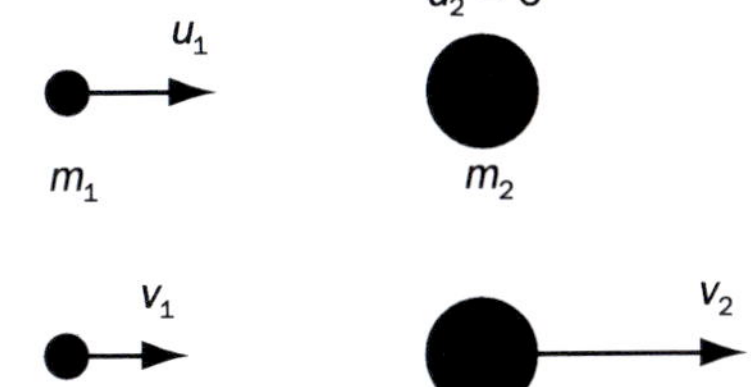

The conservation of momentum equation is the same as before:

$$m_1u_1 + 0 = m_1v_1 + m_2v_2 \quad (1)$$

The conservation of kinetic energy equation must now be modified as follows:

$$\frac{1}{2}m_1u_1^2 + 0 = \frac{1}{2}m_1v_1^2 + \frac{1}{2}m_2v_2^2 + \textit{energy lost} \quad (2)$$

Completely inelastic collisions

- Consider the same as above except that, after the collision, the two objects stick together and move with a common velocity.

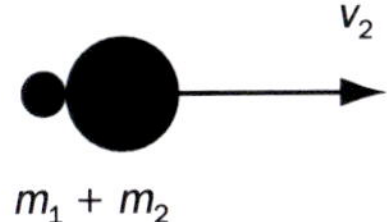

Its value is then given by the equation $m_1u_1 + 0 = (m_1 + m_2)v_2$, so that

$$v_2 = \frac{(m_1)}{(m_1 + m_2)} \ u_1.$$

Once v_2 has been calculated, it is possible to calculate the energy lost in the collision, which is given by the equation $\frac{1}{2}(m_1 + m_2)v_2^2 - \frac{1}{2}m_1u_1^2$.

- If both objects are moving before the collision, the conservation of momentum equation contains an extra component. Remember that momentum is a vector quantity and the correct signs must be used.

- Conservation of momentum in this case is $m_1u_1 - m_2u_2 = (m_1 + m_2)v_2$. v_2 would then be positive or negative, depending upon the masses and velocities of the colliding objects.

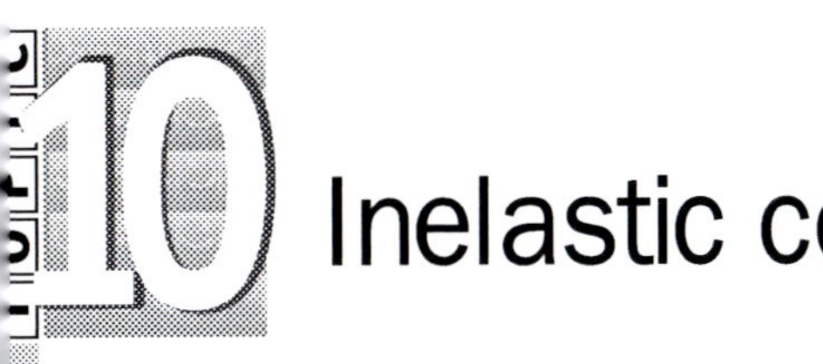

Inelastic collisions

QUESTIONS

1 An object of mass 2 kg moving with a velocity of 4 m s^{-1} collides head-on with a stationary object of mass 6 kg. After colliding, they move off together with a common velocity. What is the common velocity and how much energy is lost in the collision?

2 An object of mass 2 kg moving with a velocity of 4 m s^{-1} collides head-on with an object of mass 6 kg moving towards it with a velocity of 8 m s^{-1}. After colliding, they move off together with a common velocity. What is the common velocity and how much energy is lost in the collision?

3 A car of mass 640 kg is stationary at a road junction with its brakes off and is hit from behind by a second car of mass 950 kg travelling at 30 m s^{-1}. After the collision, they remain locked together. What would their common velocity be after the impact and how much energy is lost by the colliding car?

4 As a result of faulty signalling, two locomotives are directed on to the same track. One is travelling at 90 km h^{-1} and has a mass of 20 metric tons (1 metric ton = 1000 kg), and the other, travelling in the opposite direction, has a mass of 32 metric tons and a velocity of 60 km h^{-1}. When they collide they stick together. What is their common velocity after the collision and their resultant kinetic energy?

5 A bullet of mass 15 g is shot into a block of wood of mass 4 kg. After the collision, the kinetic energy of the block of wood and the bullet is found to be 10.6 J. Calculate the speed of the bullet.

SECTION
2
Electricity

Fields

BACKGROUND INFORMATION

Depending on the order in which topics have been taught, some parts of this lesson may contain revision work. The lesson could be left and used as revision after each of the fields has been dealt with separately.

The concept of a field is something that is used a great deal in physics. It is necessary to deal with three types of fields, namely *gravitation fields*, *electric fields* and *magnetic fields*. In each case the existence of a field is determined by the fact that a force is exerted on something when placed in the field.

- In the case of a gravitation field, a force is exerted on any mass placed in the field.
- In the case of an electric field, a force is exerted on any charge placed in the field.
- In the case of a magnetic field, a force is exerted on any electric-current-carrying conductor placed in the field.

Every field is determined by a force that acts and, therefore, all fields have direction. All fields are vectors.

Once these facts are established, it follows that:
(1) *Gravitation field strength* (g) at a point is defined as the gravitation force per unit mass on a mass at the point, i.e.

$$\textit{gravitation field strength} = \frac{\textit{gravitation force}}{\textit{mass}}$$

or, in symbols,

$$g = \frac{F}{m} \quad \text{giving} \quad F = mg$$

In SI units, the unit of gravitation field strength is newtons per kilogram, N kg^{-1}. For a person on the Earth, the gravitation field strength is approximately 9.8 N kg^{-1}. For a point as far from the Earth as the Moon, the gravitation field strength of the Earth has fallen to 0.00272 N kg^{-1}.

(2) *Electric field strength* (E) at a point is defined as the electric force per unit positive charge on a charge at the point, i.e.

$$\textit{electric field strength} = \frac{\textit{electric force}}{\textit{charge}}$$

or, in symbols,

$$E = \frac{F}{Q} \quad \text{giving} \quad F = EQ$$

One difference between the definition of electric field and gravitation field is the need to indicate that the force is that on a positive charge. The force on a negative charge would be in the opposite direction. This problem does not arise with gravitation field, since mass always attracts mass, so there is no ambiguity in its direction.

Fields

(3) *Magnetic field strength, also called magnetic flux density* (B) at a point is defined as the magnetic force per unit current in a wire at the point per unit length of the wire when the wire is at right angles to the field, i.e.

$$\textit{magnetic field strength} = \frac{\textit{magnetic force}}{\textit{current} \times \textit{length of wire}}$$

or, in symbols,

$$B = \frac{F}{Il} \quad \text{giving} \quad F = BIl$$

The unit of B, the N A^{-1} m^{-1}, is called the tesla (T).

Note that further complications arise with magnetic field, since the current-carrying conductor must have some length, and increased length results in a proportional increase in force. The wire must also have a direction of orientation, something that it was not necessary to worry about with a point mass when defining gravitation field or a point charge when defining electric field. If the wire is at any angle θ to the field, the equation becomes $F = BIl \sin \theta$.

LESSON POINTERS

Students will almost certainly have heard of magnetic fields and probably gravitation fields, but electric fields will probably be new to them. It is suggested that, for consistency, 'al' is not added to the field names, i.e. not 'magnetical', 'electrical' or 'gravitational'. It is worthwhile pointing out that this use of the word 'field' has nothing at all to do with the fields in which cows graze, but is to do with regions of space around magnets and charges and astronomical objects such as the Earth, the Moon and the Sun. In this introduction to fields, point out that each object does have its own gravitation field. A person has a gravitation field, but it is so weak that in daily life one never notices it. Even an atom has a gravitation field, but it, too, is so weak that gravity plays no part in holding atoms together in chemical bonding or in keeping solids rigid.

Most fields are three-dimensional. When students have seen or plotted magnetic fields, they have usually plotted them on a piece of paper, which gives the impression that the field is two-dimensional. They need to realise that the field does reach above and below the magnet as well as to its left and right. It should be emphasised that the field strength is a property of the region where the field is, and not a property of the object placed in the field.

The three different types of field normally dealt with at A-level should not be thought of as the only fields. Mention may be made of nuclear forces and, hence, fields within the nucleus, which are not the same as the other three.

Fields

ANSWERS TO WORKSHEET

1 1970 N (to 3 sig. figs)

2 2.5 N C^{-1}

3 240 μN

4 (i) 2.5×10^{-12}
(ii) 1.56×10^{7} N C^{-1}

5 0.144 N

6 0.63 T

7 1.08 N. When changed to 1.08 N in the opposite direction, the alteration in force is very small in comparison with the weight of the cable.

1 Fields

BASIC FACTS

- Gravitation field strength (g) at a point is defined as the gravitation force per unit mass on a mass at the point, i.e.

$$\textit{gravitation field strength} = \frac{\textit{gravitation force}}{\textit{mass}}$$

or, in symbols,

$$g = \frac{F}{m} \quad \text{giving} \quad F = mg$$

- Electric field strength (E) at a point is defined as the electric force per unit positive charge on a charge at the point, i.e.

$$\textit{electric field strength} = \frac{\textit{electric force}}{\textit{charge}}$$

or, in symbols,

$$E = \frac{F}{Q} \quad \text{giving} \quad F = EQ$$

- Magnetic field strength (B) at a point is defined as the magnetic force per unit current in a wire at the point per unit length of the wire when the wire is at right angles to the field, i.e.

$$\textit{magnetic field strength} = \frac{\textit{magnetic force}}{\textit{current} \times \textit{length of wire}}$$

or, in symbols,

$$B = \frac{F}{Il} \quad \text{giving} \quad F = BIl$$

- If there is an angle θ between the directions of the current and the field, the equation becomes

$$F = BIl \sin \theta$$

- The unit of B, the N A^{-1} m^{-1}, is called the tesla (T).

Fields

QUESTIONS

1 At a place where the Earth's gravitation field strength is 9.83 N kg^{-1}, what is the value of the gravitation force acting on a mass of 200 kg?

2 What is the value of the Earth's gravitation field strength at a place where the gravitation force on a mass of 200 kg is 500 N?

3 The value of the Earth's electric field has the surprisingly high value of 300 N C^{-1}. What is the electric force on a raindrop in a thundercloud if it has a charge of 0.80 μC?

4 An electron in a cathode ray tube has a force on it that gives it the enormous acceleration of 2.8×10^{18} m s^{-2}. Calculate
(a) the force on the electron
(b) the electric field strength of the electric field that causes this acceleration.
(Mass of electron = 9.1×10^{-31} kg; charge on electron = 1.6×10^{-19} C.)

5 A wire in an electric motor is 6.0 cm long and is at right angles to a magnetic field of field strength 0.080 T. What force is exerted on the wire when it carries a current of 3.0 A?

6 What is the magnetic field strength if a wire of length 4.0 cm, when placed at right angles to the field, experiences a magnetic force of 0.63 N when the current in the wire is 25 A?

7 An overhead power cable is at right angles to the magnetic field of the Earth of 1.8×10^{-5} T and carries a current of 200 A. Find the force exerted on a 300 m length of cable. Explain why no movement of the cable is likely when the current, which is a.c., changes to 200 A in the opposite direction.

Force between point charges

BACKGROUND INFORMATION

Electrical charge has been known about for thousands of years and, a long time ago and quite arbitrarily, the static charge acquired by glass when it was rubbed with silk was called a positive charge. The charge acquired by amber when rubbed with fur was called a negative charge. Only two types of charge are known, and they are still called positive and negative. The word 'electricity' is derived from the Greek word for amber.

The force between like charges is always one of repulsion, and the force between unlike charges is always one of attraction. In other words, positive charge repels positive charge; negative charge repels negative charge; positive charge attracts negative charge.

The magnitude of the force between two point charges is proportional to the two charges and inversely proportional to the square of the distance between them. Consider the arrangement of two point charges $+Q_1$ and $+Q_2$ separated by a distance r, as shown in Figure 1.

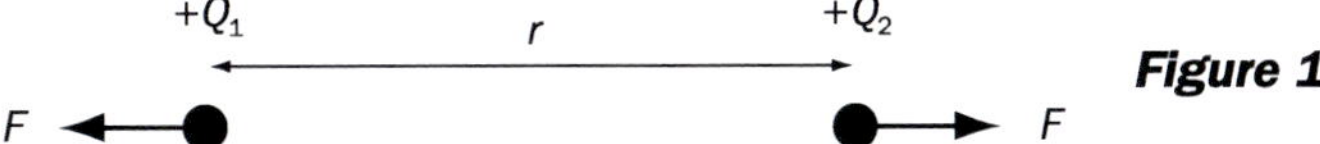

Figure 1

Using Newton's third law, the force F that charge $+Q_1$ exerts on charge $+Q_2$, must be equal and opposite to the force F that charge $+Q_2$ exerts on charge $+Q_1$. This gives

$$F \propto \frac{Q_1Q_2}{r^2}$$

Use of a constant of proportionality (k) makes possible the removal of the proportional sign and the insertion of an equals sign, giving

$$F = \frac{kQ_1Q_2}{r^2}$$

When determined practically, the value of this constant, k, is found to have the very high value of 8.99×10^9 N m^2 C^{-2}. The practical implication of the very high value is that even small charges have strong forces acting between them. It is the electric forces between atoms rather than gravitation forces that hold matter together.

Partly for historical reasons and partly to simplify formulae derived from the above equation, it is not usual to write the constant k as we have done above. It is normally written with

$$k = \frac{1}{4\pi\varepsilon_0}$$

Since the value of k is known, the value of ε_0 is given by

$$\varepsilon_0 = \frac{1}{4\pi k} = \frac{1}{4 \times \pi \times 8.99 \times 10^9} = 8.85 \times 10^{-12}\ \text{C}^2\ \text{m}^{-2}\ \text{N}^{-1}$$

ε_0 is a fundamental constant and is called the permittivity of free space. Because the constant is used a great deal when dealing with the theory of capacitors, the equivalent unit, farad per metre (F m^{-1}), is often used, giving

permittivity of free space, $\varepsilon_0 = 8.85 \times 10^{-12}$ F m^{-1}

Force between point charges

TEACHER NOTES

LESSON POINTERS

This is a difficult topic to teach, largely because it seems to have little connection with everyday life, and examples tend to be very artificial. It is not made easier by the way in which the constant of proportionality is treated. However, students need to be reminded of the reasons for this apparently clumsy introduction of ε_0 when later, simpler, formulae occur. The point of this rationalisation is to make situations in which there is a spherical field, such as the one considered, have a 4π term in any formula. Those involving a cylindrical field will then have a 2π term in any formula, and there will not be a π term in situations involving a uniform field, such as in the formula for the capacitance of a parallel plate capacitor.

ANSWERS TO WORKSHEET

1 Repulsion of 1.8×10^4 N. This is a huge value, so free charges in coulombs are impracticable.

2 Attraction of 4.1×10^{-8} N

3 Repulsion of 1.6 N. A very large force on such a small mass.

4 28.3°. The charges on the balls must be considered as point charges. Note that, although the charges are different, the forces on each ball must be the same and so, since the weights of the balls are the same, the diagram is symmetrical. The angle is found from the fact that the resultant of the force of repulsion and the weight must be in line with the string.

Force between point charges

BASIC FACTS

- The magnitude of the force between two point charges is proportional to the two charges and inversely proportional to the square of the distance between them.

Consider two point charges $+Q_1$ and $+Q_2$ separated by a distance r.

$$F \propto \frac{Q_1Q_2}{r^2}$$

Use of a constant of proportionality (k) makes possible the removal of the proportional sign and the insertion of an equals sign, giving

$$F = \frac{kQ_1Q_2}{r^2}$$

- k is normally written in the following way as, rather surprisingly, it simplifies a later theory

$$k = \frac{1}{4\pi\varepsilon_0}$$

Since the value of k is known, the value of ε_0 is given by

$$\varepsilon_0 = \frac{1}{4\pi k} = \frac{1}{4 \times \pi \times 8.99 \times 10^9} = 8.85 \times 10^{-12}\ \mathrm{C^2\ m^{-2}\ N^{-1}}$$

- ε_0 is a fundamental constant and is called the permittivity of free space. Because the constant is used a great deal when dealing with the theory of capacitors, the equivalent unit, farad per metre ($\mathrm{F\ m^{-1}}$) is often used, giving

permittivity of free space, $\varepsilon_0 = 8.85 \times 10^{-12}\ \mathrm{F\ m^{-1}}$

QUESTIONS

1 Determine the force between one point charge of 1.0 C and another point charge of 2.0 C when the charges are placed 1000 m apart in a vacuum. Why is this question unrealistic?

2 Determine the force between a chlorine ion and a sodium ion when separated by a distance of 2.5×10^{-10} m. (A sodium ion has a charge of $+1.60 \times 10^{-19}$ C and a chlorine ion a charge of 1.60×10^{-19} C.)

3 Determine the force between an alpha particle (charge $+3.2 \times 10^{-19}$ C) and a gold nucleus (charge $+9.6 \times 10^{-18}$ C), when a distance of 1.3×10^{-13} m separates them.

4 Two evenly charged balls, each of weight 0.025 N, are suspended as shown in the figure and come to rest with their centres a distance 0.20 m apart when the left-hand ball carries a charge of +600 nC and the right-hand ball carries a charge of +100 nC.

Calculate the angle each string makes with the vertical. What assumption did you have to make in order to be able to start this question?
(1 nC = 1 nanocoulomb = 1×10^{-9} C.)

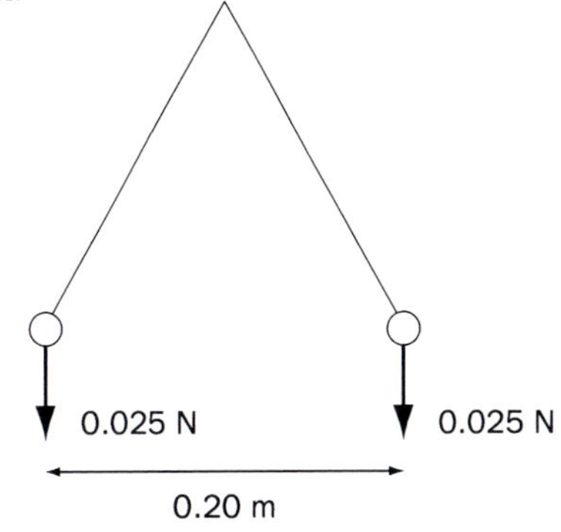

Electric field

BACKGROUND INFORMATION

In the same way that field lines can be drawn around magnets, so can field lines be drawn around charges. In the case of magnets, the field lines show the strength and direction of the magnetic field. The strength is shown by the concentration of field lines; near the poles of a magnet there should be a cluster of field lines. Near an electric charge there is an electric field and, again, the concentration of the field lines should show the strength of the electric field, and there should be an arrow on a field line to show its direction.

The definition of electric field, which usually has the symbol *E*, is in terms of the force per unit positive charge placed in the field.

Electric field strength (*E*) at a point is defined as the electric force per unit positive charge on a positive charge at the point.

The force on a negative charge would be in the opposite direction, i.e.

$$\text{electric field strength} = \frac{\text{electric force}}{\text{charge}}$$

or, in symbols,

$$E = \frac{F}{Q} \text{ giving } F = EQ$$

Since electric field is defined as a force per unit positive charge, it must follow that electric field strength is a vector.

In order to calculate the magnetic field strength at a particular point, use is made of the force of attraction/repulsion between charges. A unit positive charge is imagined, placed at the point where the electric field strength is required. In practice, of course, this would be impossible, as a charge of one coulomb is impossibly large. In reality, the force on a much smaller point charge would be used, and then the measurement would be multiplied by the correct multiplying factor.

If the resultant of two or more fields is required, addition of fields must be done vectorially using a parallelogram of forces. At this stage, it would be sensible to do one or more of the questions on the Student Worksheet as worked examples.

LESSON POINTERS

There is a good opportunity at this stage to get students to practise calculating fields, to revise vector addition and to draw field patterns. Depending on their mathematical ability, students can either calculate the resultant field, using the trigonometry of a triangle, or they can make a scale drawing. When drawing field patterns, they should be encouraged not to think of fields as 'fluffy' line drawings but to treat a field as a definite and accurately determinable quantity. It is worthwhile to have a rectangular box drawn around the field drawing and to draw the whole field within this box. This will be less likely to lead to drawings that leave large blank spaces where fields ought to be shown, and will be more likely to show a strong field with many field lines and a weak field with

Electric field

few field lines. The three-dimensional nature of electric fields is difficult to show in a drawing, so it should be mentioned several times that normally sketches of fields merely show the field in one plane only.

ANSWERS TO WORKSHEET

1 3360 N C^{-1}

2 0.024 N

3 (a) 1.17×10^{10} N C^{-1}
(b)

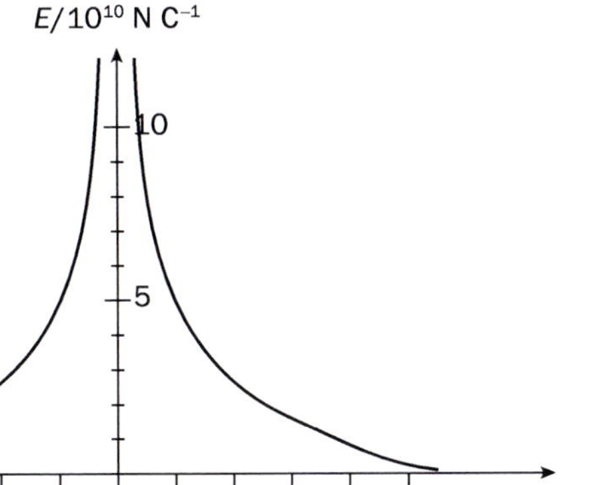

(c)

4 (A) 10.8×10^6 N C^{-1}
(B) 3.8×10^6 N C^{-1}
(C) 1.35×10^6 N C^{-1}
(D) 2.0×10^6 N C^{-1}
Field pattern shown rotated 90° in 5b

5

3 Electric field

BASIC FACTS

- Electric field strength (E) at a point is defined as the electric force per unit positive charge on a charge at the point:

$$E = \frac{\textit{electric force}}{\textit{charge}} = \frac{F}{Q}$$

- One difference between the definitions of electric field and gravitation field is the need to indicate that the force is that on a positive charge. The force on a negative charge would be in the opposite direction.
- Electric field strength can be calculated by using the equation giving the force of repulsion/attraction between the charge or charges causing the field and an imagined unit positive charge placed at the point where the field is required.
- Electric field strength is a vector, so vector addition must be used if adding field strengths.

QUESTIONS

1 What is the electric field strength at a point if a charge of 25 nC experiences a force of 84 μN when placed at the point?

2 What force is exerted on a charge of 300 nC when it is placed at a point where the electric field strength is 80 000 N C^{-1}?

3 (a) Calculate the value of the electric field strength at a distance of 3.5×10^{-10} m from a charge of $+1.6 \times 10^{-19}$ C.

(b) Sketch a graph showing how the electric field strength varies with distance x from a charge of $+1.6 \times 10^{-19}$ C. Use both positive and negative values of x.

(c) Draw an electric field diagram showing the field near a positive charge. How would the diagram differ if the field were to be drawn near a negative charge of the same magnitude?

4 A charge of +6.0 μC is placed at a distance of 0.20 m from a charge of −6.0 μC, as shown in the diagram. Calculate the magnitude and direction of the resultant field at points A, B, C and D.

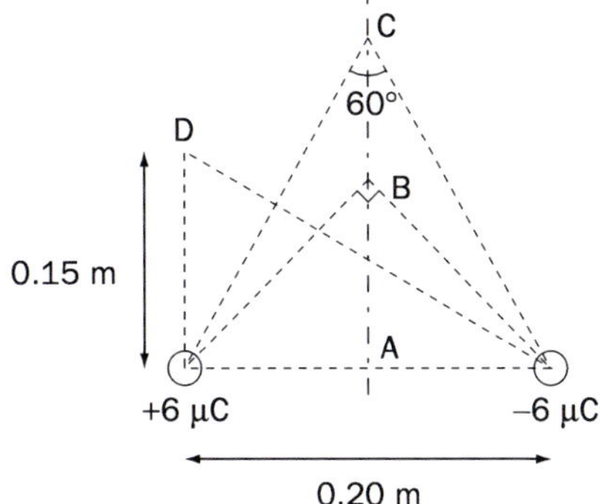

Draw a diagram showing the field pattern in the space around the charges.

3 Electric field

5 Draw diagrams showing the shape of the electric field for the following arrangements.

(a)

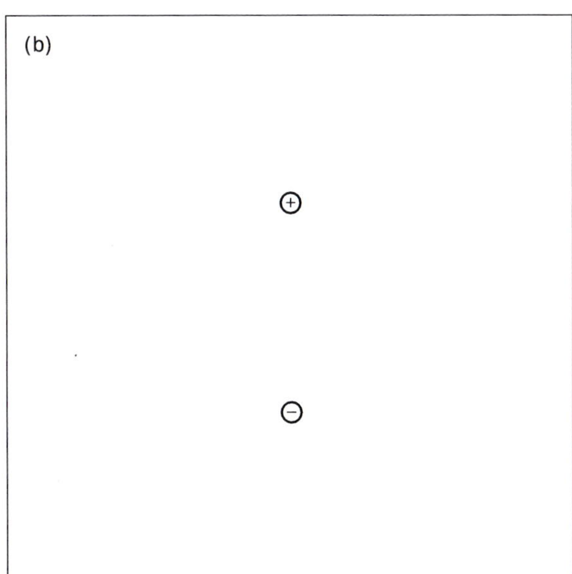

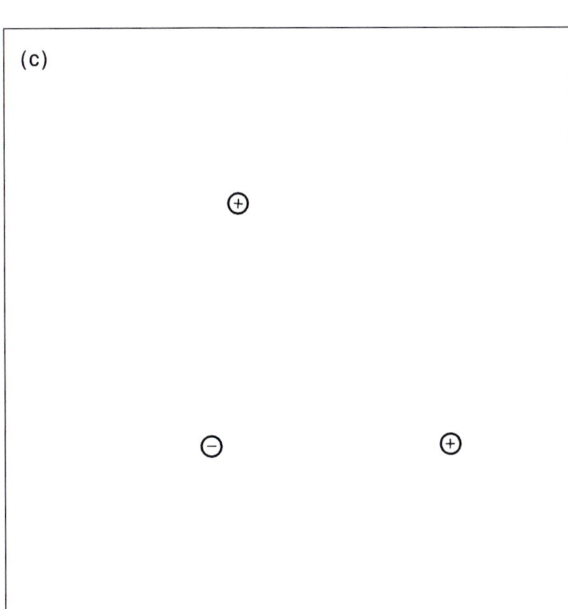

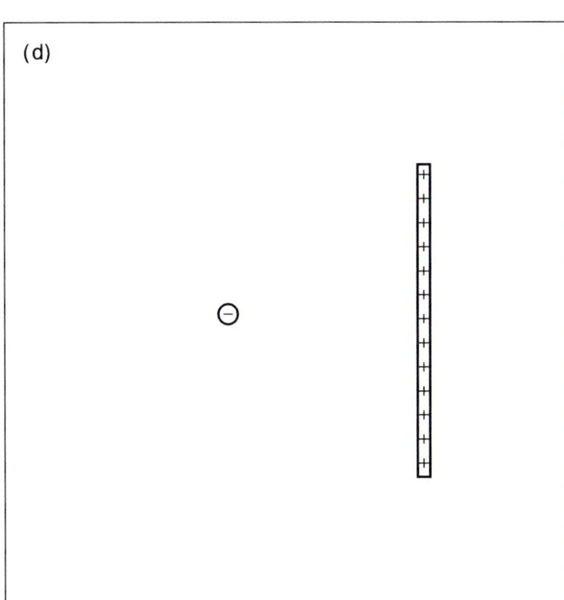

Potential gradient

BACKGROUND INFORMATION

Consider first two plates of equal area, parallel and close to one another, a distance x apart, with the top one charged positively and the bottom one charged with an equal but negative charge. The arrangement is shown in Figure 1, together with the electric field between the plates.

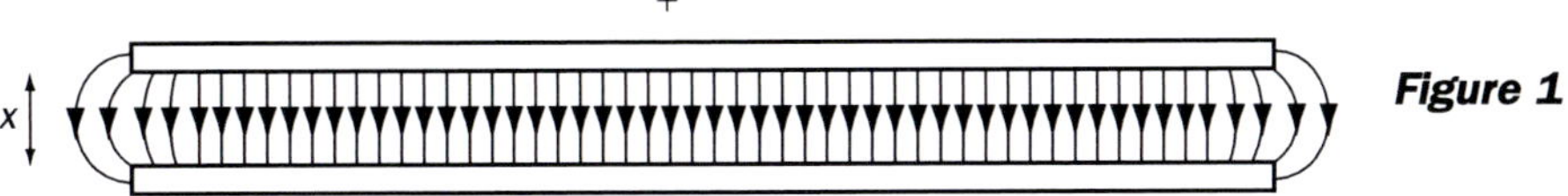

Figure 1

It can be seen that the field is uniform except near the edges of the plates. Now consider a charge $+Q$ near to the centre of the plates, so that it is in the uniform part of the field, as shown in Figure 2.

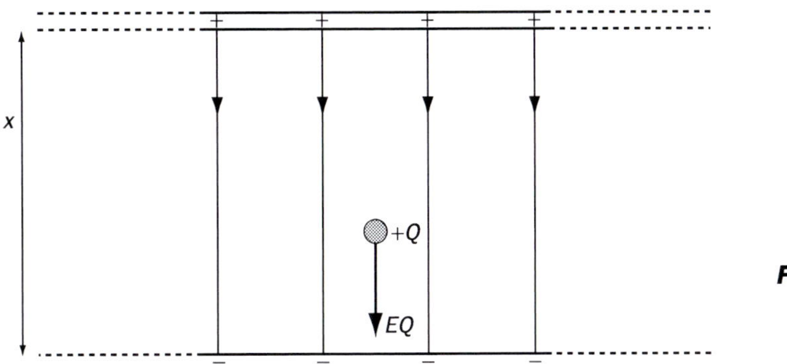

Figure 2

From the definition of electric field, the force on charge $+Q$ is EQ downwards. If, therefore, the charge is to be moved at a constant velocity from the bottom plate to the top plate, some work has to be done by an equal external force. The amount of work done by this external force is given by

work done = force × distance moved in the direction of the force

$$W = EQ \times x$$

When we dealt with current electricity, the physical quantity potential difference, V, between two points was introduced as the work done in moving unit charge between the two points. If this definition is used here, we find that

$$V = \frac{W}{Q} = \frac{EQx}{Q} = Ex$$

This gives $E = \frac{V}{x}$ or that the electric field is the potential gradient.

Two equations for electric field are, therefore, the defining equation

$$E = \frac{F}{Q} \text{ or } E = \frac{V}{x}$$

Electric field is often measured in volts per metre, but newtons per coulomb is an equally correct alternative.

Potential gradient

LESSON POINTERS

This is another topic which many students find difficult. One reason for this is that many, even by this stage, still have only a sketchy understanding of potential difference, so it is worthwhile starting this topic by carrying out a revision of this term and its unit, the volt. Another complication here is that very many questions involving electric field require students to appreciate that electric field is both the force acting per unit charge and also the potential gradient. Students often seem locked into one of these ways of considering electric field but, without both of them, they find many simple questions very difficult. In particular, they should be encouraged to work from considerations of energy rather than force. This often makes many problems almost trivial.

The details given in this topic are for a uniform field. The same principles can be used for a non-uniform field, but then calculus is required to get the electric field as the potential gradient.

$$E = -\frac{dV}{dx}$$

The minus sign in this equation arises from the fact that V rises as x decreases. The convention that potential is zero at infinity could also be introduced at this stage for the more able students, together with the use of the term potential rather than potential gradient or potential difference.

Other points to be careful with in this lesson are

(i) the weight of the charge $+Q$ is being ignored. This is because gravitational forces in these problems are infinitesimally small compared with electric forces. The plates may be considered as being side by side if this is thought to be preferable;

(ii) the movement of the charge at constant velocity causes problems with some students. 'How can it move if there is no resultant force on it?' While this could lead to a discussion of Newton's laws, this is unlikely to be helpful in the middle of this lesson, as it can involve a considerable amount of side-tracking. When this problem is first met, it is probably better simply to state that there is no acceleration with zero force, so movement at constant velocity can be correctly considered over the whole distance x.

Potential gradient

ANSWERS TO WORKSHEET

1 Both are kg m s^{-3} A^{-1} in base units

2 (a) 0.168 N
(b) 3.86×10^{-4} J
(c) 64.4 V increase

3 1.93×10^{7} m s^{-1}

4 (a) 13 600 V m^{-1}
(b) 2.18×10^{-15} N vertically upwards
(c) 2.9×10^{7} m s^{-1}. This is a crucial step: the horizontal velocity is unchanged since there is no horizontal force.
(d) 1.66×10^{-9} s
(e) 2.40×10^{15} m s^{-2}
(f) 3.96×10^{6} m s^{-1}
(g) $\tan \theta = v_v/v_h = 0.102$; $\theta = 5.8°$
(h) 39.2 m s^{-1}

5 (a) increase
(b) decrease
(c) increase
(d) decrease

4 Potential gradient

BASIC FACTS

- For a uniform electric field, the potential difference V between two points can be related to the electric field in the following way.

  $$V = \frac{W}{Q} = \frac{EQx}{Q} = Ex$$

  This gives $E = \frac{V}{x}$ or that the electric field is the potential gradient.
- Two equations for electric field are therefore the defining equation

  $$E = \frac{F}{Q} \text{ or } E = \frac{V}{x}$$

- The unit of electric field is therefore either newtons per coulomb ($N\,C^{-1}$) or volts per metre ($V\,m^{-1}$). In many problems it is necessary to use both of these equations; if you remember only one of them, easy questions can become difficult or impossible.

QUESTIONS

1 Show that the newton per coulomb is the same unit as the volt per metre.

2 A charge of +6.0 μC is moved 2.3 mm in the opposite direction to an electric field of magnitude 2.8×10^4 N C^{-1}.
(a) Calculate the external force on the charge.
(b) How much work is done on the charge?
(c) By how much does the potential of the charge change? Is the change an increase or a decrease?

3 Calculate the speed that an electron acquires if it is accelerated from rest for a distance of 280 mm by an electric field of field strength 3.8×10^5 V m^{-1}.

Potential gradient

4 An electron travelling with a velocity of 2.9×10^7 m s^{-1} enters the space between the two Y-plates of an oscilloscope, as shown in the diagram. The plates are 48 mm long and 22 mm apart, the top plate is at a potential of +150 V and the bottom plate is at a potential of –150 V.

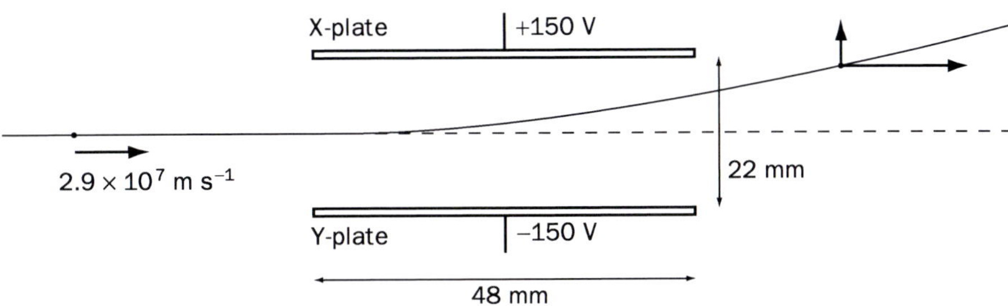

Calculate, ignoring edge effects throughout,
(a) the electric field between the plates
(b) the force (magnitude and direction) on the electron while it is between the plates
(c) the horizontal component of the velocity of the electron as it leaves the plates
(d) the time the electron is between the plates
(e) the vertical acceleration of the electron while between the plates
(f) the vertical component of the velocity of the electron as it leaves the plates
(g) the direction of the electron's velocity as it leaves the plates
(h) the magnitude of the electron's velocity as it leaves the plates.

5 Without working out values, deduce whether the angular deflection of the electron in question 4 increases, decreases or remains constant when, separately,
(a) the potential difference between the plates is increased
(b) the separation of the plates is increased
(c) the length of the plates is increased
(d) the initial velocity of the electron is increased.

Capacitance

BACKGROUND INFORMATION

As we saw in the last topic, when a pair of parallel plates have equal but opposite charges on them, there is an electric field in the space between the plates and there is a potential difference between the plates. Experimentally and theoretically, it can be shown that the charge on each plate is proportional to the potential difference between them. The ratio of the charge to the potential difference is called *capacitance*.

Capacitance is defined as the charge on each plate per unit potential difference, resulting in the equation

$$C = \frac{Q}{V}$$

The SI unit of capacitance must therefore be the coulomb per volt ($C\ V^{-1}$). One coulomb per volt is called a farad (F). This is a very large capacitance so, frequently, the μF (microfarad) = 10^{-6} F, the nF (nanofarad) = 10^{-9} F and the pF (picofarad) = 10^{-12} F are used.

COMBINATIONS OF CAPACITORS IN PARALLEL

Consider the circuit shown in Figure 1, with two capacitors C_1 and C_2 in parallel with a potential difference V across them.

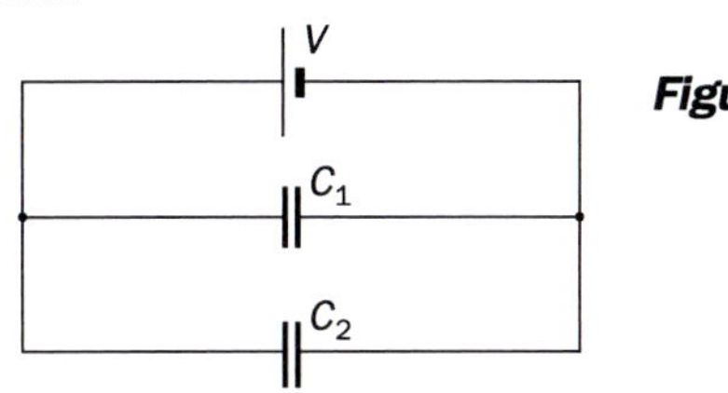

Figure 1

The charge on each capacitor and the total charge Q are given by

$$Q_1 = C_1V, \quad Q_2 = C_2V \quad \text{and, therefore,} \quad Q = C_1V + C_2V$$

Since the total charge Q = *total capacitance* × *potential difference*, and since V then cancels throughout the equation, we get $C = C_1 + C_2$.

COMBINATIONS OF CAPACITORS IN SERIES

Now consider two capacitors in series as shown in Figure 2.

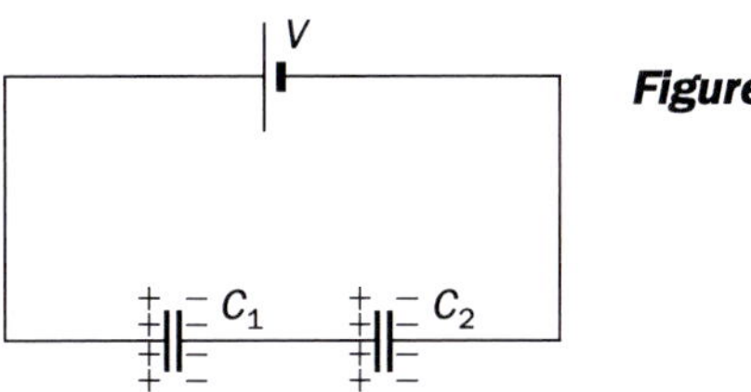

Figure 2

The charge that leaves the power supply distributes itself on the plates in the way shown in the diagram. The charge on each capacitor must be the same and this will also equal the charge Q from the supply. This gives the potential difference across each capacitor as

$$V_1 = \frac{Q}{C_1} \qquad V_2 = \frac{Q}{C_2} \qquad \therefore V = \frac{Q}{C_1} + \frac{Q}{C_2} = \frac{Q}{C}$$

Cancelling Q gives

$$\frac{1}{C_1} + \frac{1}{C_2} = \frac{1}{C}$$

5 Capacitance

LESSON POINTERS

This topic should be backed up by setting the students a large number of examples to work through, so that they gain familiarity with the topic. One problem that arises as soon as the definition is given concerns the charge on a capacitor. If it is a parallel plate capacitor, there will be a charge $+Q$ on one plate and a charge $-Q$ on the other. Under these circumstances, the capacitor is said to have a charge Q (i.e. not zero and not $2Q$).

COMBINATIONS OF CAPACITORS IN PARALLEL

This normally proves to be quite straightforward, even though the result is opposite to the result for resistors.

COMBINATIONS OF CAPACITORS IN SERIES

This equation, in itself, does not usually cause any problem, but there is one place in deducing the equation that is a source of much confusion. This concerns the charge on capacitors in series. That these charges are equal can be explained easily, if necessary by imagining that there are ammeters at several points around the circuit. The ammeters all read identical currents at every instant, so the charge that passes any point must be the same. This clearly implies that the charge that has been supplied by the power source must also be equal to the charge on each capacitor. So far, so good. The problem comes when the student is left to do a problem on his or her own. Then, if they get 300 nC on one capacitor and 300 nC on the next capacitor, they almost automatically assume that the total charge supplied is 600 nC. It is probably necessary for them to make this mistake a few times so that, hopefully, they learn from the mistake.

ANSWERS TO WORKSHEET

1 (a) 30 μF
(b) +1800 μC; –1800 μC

2 (a) 56.4 μF; 37.6 μF
(b) 677 μC; 451 μC

3 30 V

4 (a) 10 μF
(b) 1200 pF
(c) 1000 μF

5 4.0, 18.0, 9.0, 10.0 in μF; 20, 6.7, 13.3, 20.0 in V; 80, 120, 120, 200 in μC

Capacitance

BASIC FACTS

- Capacitance C is defined as the charge Q on each plate per unit potential difference, resulting in the equation

$$C = \frac{Q}{V}$$

The SI unit of capacitance must therefore be the coulomb per volt ($C\ V^{-1}$). One coulomb per volt is called a farad (F). This is a very large capacitance, so smaller units are often used.
1 μF (microfarad) = 10^{-6} F
1 nF (nanofarad) = 10^{-9} F
1 pF (picofarad) = 10^{-12} F

- For capacitors in parallel, the total capacitance is given by $C = C_1 + C_2 + \ldots$.
- For capacitors in series, the total capacitance is given by

$$\frac{1}{C} = \frac{1}{C_1} + \frac{1}{C_2} + \ldots$$

QUESTIONS

1 A capacitor has a potential difference of 20.0 V across its plates. The charge on one plate is +600 μC and on the other plate is –600 μC. Calculate
(a) the value of the capacitance of the capacitor
(b) the charge on each plate when the potential difference between the plates is changed to 60.0 V.

2 A capacitor is marked 47 μF ± 20%.
(a) What are the maximum and minimum values for the capacitance of the capacitor?
(b) What are the maximum and minimum values for the charge stored by the capacitor when it has a potential difference of 12 V across its plates?

3 What potential difference is necessary across a 0.022 μF capacitor if the charge on the capacitor is to be 0.66 μC?

4 Calculate the total capacitance of the following arrangement of capacitors

(a)

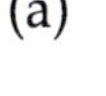

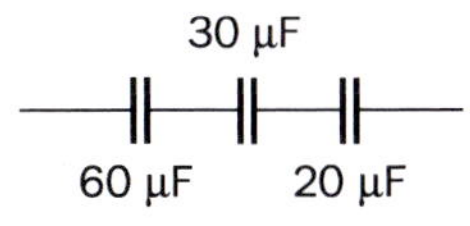

(b)

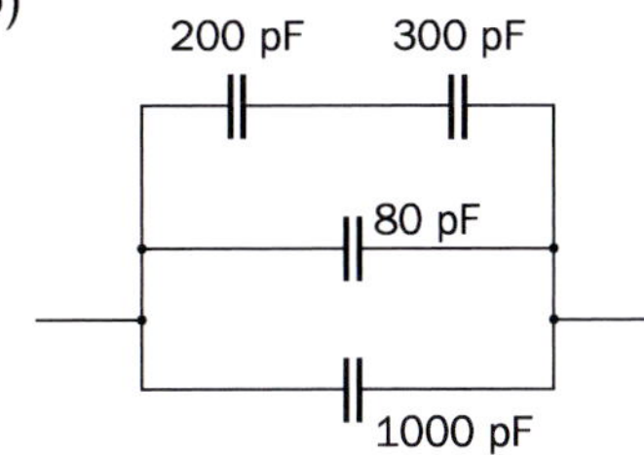

(c)

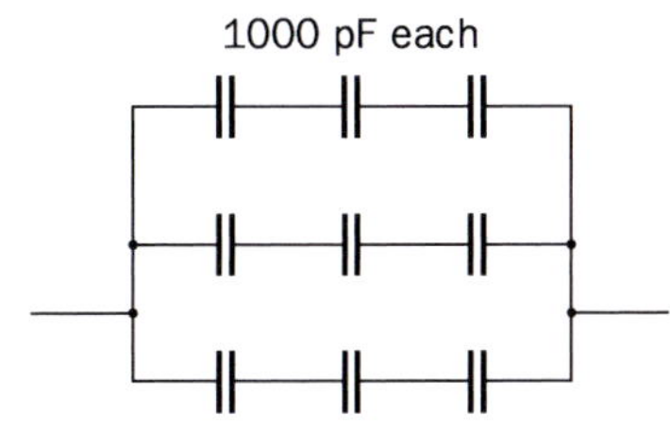

Capacitance

5 Complete the table to show the charge, potential difference and capacitance for each capacitor in the following circuit. Include the figures for the whole circuit in the column labelled 'total'.

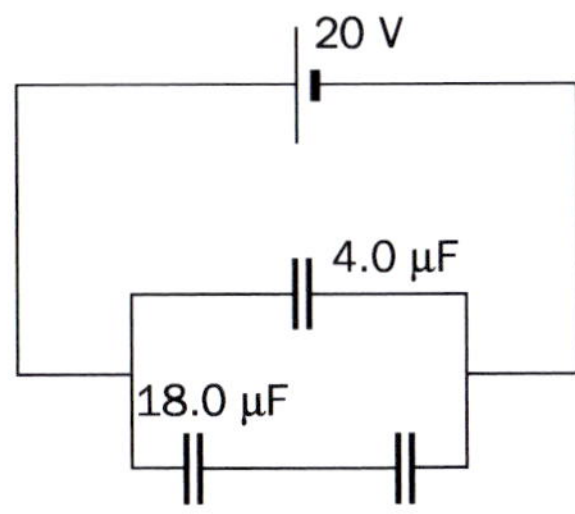

	—	—	**Unlabelled capacitor**	**Total**
Capacitance	4.0 μF	18.0 μF		
Potential difference				20.0 V
Charge		120 μC		

Energy stored in a charged capacitor

BACKGROUND INFORMATION

When a capacitor is charged by gradually increasing the potential difference across it, the charge on the capacitor increases linearly with the potential difference (see Figure 1).

It becomes increasingly difficult to add additional charge to an already charged capacitor because of the repulsion effect of the charge already on the plates. More work has to be done to add 1 μC of charge to a capacitor carrying a charge of 50 μC than to one carrying a charge of only 10 μC. This implies that if a charge of, say, 200 μC is on a capacitor with a potential difference across it of 20 V ($1\ \text{V} = 1\ \text{J C}^{-1}$), it is not possible to say that the energy stored by the capacitor is $200\ \mu\text{C} \times 20\ \text{J C}^{-1} = 4000\ \mu\text{J}$.

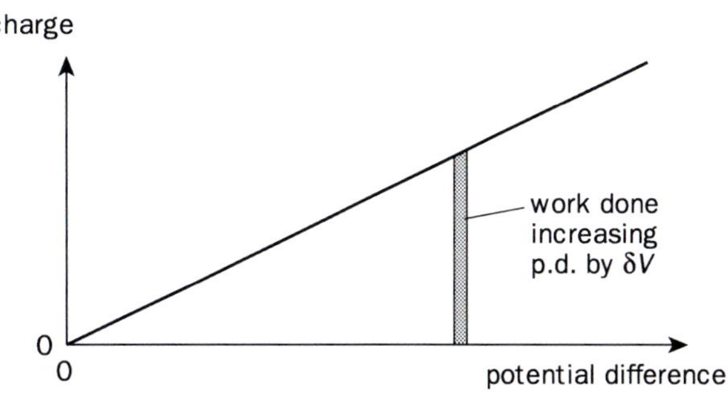

Figure 1

The energy actually stored is given by the area beneath the Q–V graph, and so will be $\frac{1}{2}QV$, or in this case $\frac{1}{2} \times 200\ \mu\text{C} \times 20\ \text{J C}^{-1} = 2000\ \mu\text{J}$.

By substitution from the equation defining capacitance, $C = Q/V$, we get

$$\textit{energy stored } E = \frac{1}{2}QV = \frac{1}{2}\frac{Q^2}{C} = \frac{1}{2}CV^2$$

Note that if C is in microfarads, Q is in microcoulombs, and V is in volts (not microvolts), E will be in microjoules. This can save a great deal of time and a fair number of careless mistakes with powers of ten.

LESSON POINTERS

While students are usually quite happy to accept that the area beneath the Q–V graph gives the work done in charging the capacitor and, hence, the energy stored by the capacitor, those who are capable of doing the necessary integration to deduce the result ought to be encouraged to do so. All candidates should be shown the graph with the area required shaded, and most should be given the opportunity to see that the area required is really the summation of a series of strips. By visualising the graph, it is to be hoped that students will remember the $\frac{1}{2}$ term, which appears in the equation and is the cause of so many mistakes on this and similar topics.

The analogy of stretching a spring a distance of 0.40 m by using a force of up to 200 N can be introduced. The work done is not $F \times x = 80$ J. The average force is half the final force, so the work done is $\frac{1}{2}Fx = \frac{1}{2} \times 200\ \text{N} \times 0.40\ \text{m} = 40$ J.

Some students will require help with the algebra to get the energy stored in terms of C, V and Q.

Energy stored in a charged capacitor

ANSWERS TO WORKSHEET

1 0.45 J

2 (a) 9.1 V
(b) 9.1 mJ

3 (a) 15 μF
(b) 600 μC on each
(c) 8.2 mJ in 22 μF capacitor and 3.8 mJ in 47 μF capacitor. (The larger value capacitor stores the smaller amount of energy.)

4 (a) 1200 μC; 7.2 mJ
(b) 4 V
(c) 0.80 mJ in 100 μF capacitor and 1.6 mJ in 200 F capacitor. Energy is lost in a spark when the second capacitor is joined to the first.

5 This capacitor stores only 125 J. A light bulb will use this energy in 2 s, so millions of these large capacitors would be required to store the hundreds of millions of joules that a power station supplies in a second or so (apart from the a.c./d.c. problem and the variable potential difference as the capacitor discharges).

Energy stored in a charged capacitor

BASIC FACTS

- The energy stored in a charged capacitor is equal to the area beneath a charge–potential difference graph.

$$\textit{energy stored } E = \tfrac{1}{2}QV = \frac{1}{2}\frac{Q^2}{C} = \tfrac{1}{2}CV^2$$

- Instead of using C, F, J and V as units here it is possible to use μC, μF, μJ and V.

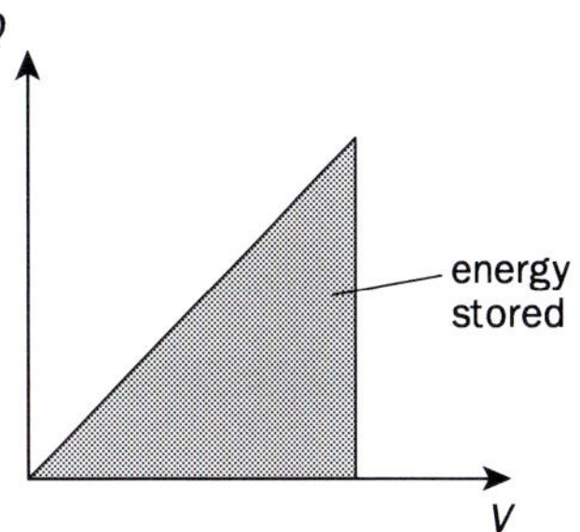

QUESTIONS

1 How much energy is stored by a 1000 μF capacitor when there is a potential difference of 30 V across it?

2 A capacitor of capacitance 220 μF stores a charge of 2000 μC. Calculate
(a) the potential difference across the capacitor
(b) the energy stored by the capacitor.

3 A capacitor with a capacitance of 47 μF is placed in series with another capacitor with a capacitance of 22 μF and a supply of 40 V is placed across them. Calculate
(a) the total capacitance
(b) the charge on each capacitor
(c) the energy stored by each capacitor.

4 A capacitor of capacitance 100 μF is charged by using a 12 V supply. The supply is then removed and a second capacitor, of capacitance 200 μF is placed across the first, so sharing the charge. Calculate
(a) the charge and energy initially stored by the 100 μF capacitor
(b) the potential difference across both capacitors when in parallel
(c) the energy stored by each capacitor when they are in parallel.
Explain why some energy seems to have disappeared.

5 A very large capacitor has a capacitance of 100 000 μF and can have a potential difference across it of 50 V without there being any damage caused to it. Use this information to show why it is that any electrical power supply company cannot realistically use capacitors to store electrical energy in order to smooth out supply and demand.

Charge and discharge of capacitors

BACKGROUND INFORMATION

When a charged capacitor is discharged through a resistor, the charge does not fall away linearly. The time taken to reach any particular fraction of the original charge depends on the capacitance of the capacitor and also on the resistance of the resistor. The larger the value of each of these quantities, the longer the time taken. It is found that the product of capacitance and resistance controls the rate of discharge. This product is called the *time constant* (usual symbol Greek tau, τ) of the circuit, i.e.

$$\tau = CR$$

During one time constant, the charge on a capacitor falls to 0.3678 of its value at the start of the time constant; during two time constants it will therefore fall to 0.3678 of 0.3678 of its initial value, that is to 0.1353 of the initial value. Continuing this sequence shows that, after 5τ, the charge on the capacitor is less than 1% of its initial value ($0.3678^5 = 0.0067$). This is shown graphically in Figure 1, together with the circuit used.

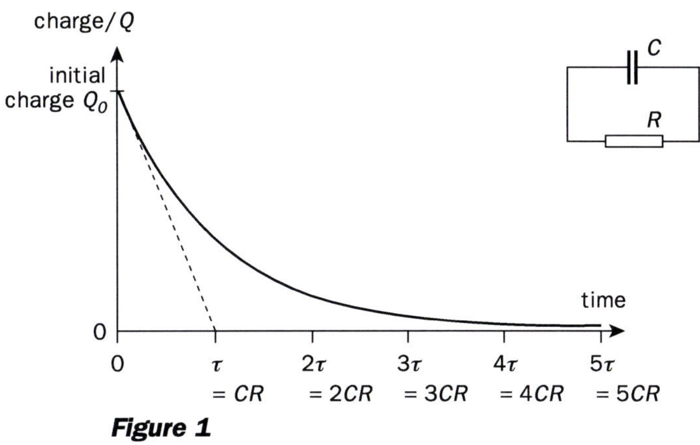

Figure 1

The dotted line on the graph is a tangent to the graph at $t = 0$. It can be seen that, if the initial rate of discharge had been maintained, the capacitor would have become totally discharged after a time equal to one time constant. Question 4 asks students to show that this is true for a particular case. Able students should be able to prove this in the general case after answering the question.

The equation of the graph in Figure 1 is

$$Q = Q_0 \mathrm{e}^{-\frac{t}{CR}}$$

From this equation it follows that the potential difference V across the capacitor, and also across the resistor, is given by

$$V = V_0 \mathrm{e}^{-\frac{t}{CR}}$$

where V_0 is the potential difference across the capacitor at time $t = 0$.

A corresponding graph is drawn in Figure 2 for a capacitor being charged through a resistor from a power supply of electromotive force (e.m.f.) V_0, together with the circuit diagram.

Charge and discharge of capacitors

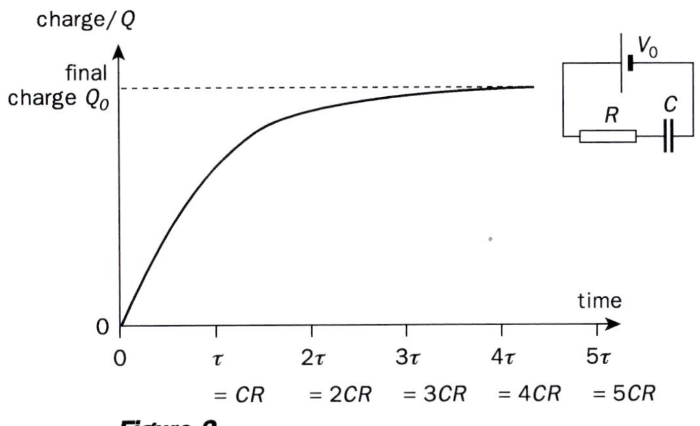

Figure 2

The equation for this Q–t graph is

$$Q = Q_0(1 - e^{-\frac{t}{CR}})$$

The corresponding potential difference–time equation is

$$V = V_0(1 - e^{-\frac{t}{CR}})$$

where V is the potential difference across the capacitor. This potential difference rises during charge, so the potential difference across the resistor must fall as the charging current falls. The potential difference across the resistor V_R is given by

$$V_R = V_0 - V = V_0 - V_0(1 - e^{-\frac{t}{CR}}) = V_0 e^{-\frac{t}{CR}}$$

LESSON POINTERS

The way in which the theory of this topic is handled will depend crucially on whether or not students have dealt with natural logarithms and the integration of $1/x$. Those students who have seen the mathematics of this topic should not find any great difficulty in proving the charge or discharge equations, though they will probably find it strange using Q and t as variables instead of x and y. However, the proof of these equations is not required for any A-level Physics candidate. Those candidates who have not done calculus will not, therefore, be at any disadvantage provided they can use the equations together with their calculators to obtain correct numerical values. It is, therefore, essential to spend some time doing some work on logarithms with candidates who have not met them before.

The first thing that students need to realise about a logarithm is that it is an index. They are all familiar with the fact that $10^3 = 1000$: now they need to be told that 3 is the logarithm of 1000 when 10 is used as a base number. Similarly since $10^4 = 10\,000$, log 10 000 = 4. They should check these and other similar calculations with their calculators. Include such expressions as $10^{-4} = 0.0001$, so log 0.0001 = –4. When they have done this, get them to work backwards using the shift 10^x button on the calculator. The next step is to deal with fractional indices. Since $\sqrt{10} = 10^{1/2}$, the logarithm of 3.16 227 766 is 0.5 and the logarithm of 31.622 776 6 is 1.5. All positive numbers have a corresponding logarithm, since all positive numbers can be expressed in index form somehow. Negative numbers and zero do not have a logarithm.

Once this principle is established using 10 as a base, then introduce the idea that any number other than 10 could also be used. You will have to introduce e (= 2.718 281 828...) as a number used for reasons not stated, for a so-called natural logarithmic base. This is marked ln on calculators, and it stands for natural logarithm, not for the word ln. The shift ln button gives e to the power x. It is useful that e^3 is almost exactly 20, so ln 20 = 3, ln 400 = 6, ln 8000 = 9 etc. In the same way that students should try out using the log/lg function, so they should now try using the ln and shift ln

Charge and discharge of capacitors

buttons, because they will certainly need them in dealing with the charge and discharge of capacitors.

A point to notice about the time constant τ is that it does not correspond to the decay constant in radioactivity. It is a pity that capacitor discharge and radioactivity, although both exponential decays, are not usually dealt with mathematically in the same way.

ANSWERS TO WORKSHEET

1 $CR = \frac{Q}{V} \times \frac{V}{I} = \frac{Q}{I} = \frac{It}{I} = t$

2 (a) 470 s
(b) 564 μC
(c) 3.8 μC
(d) 0.081 V

3 (a) 600 μC
(b) 1.8 mJ
(c) 2 s
(d) 10 s

4 (a) 90 μC
(b) 7.5 s
(c) 6.0 V
(d) 12 μA
(e) 7.5 s

5 1.39 s

6 (a) 0.90 μC
(b) 0.50 μC
(c) 5.0 V
(d) 4.0 V
(e) 1.8 μA

7 120 Ω

Charge and discharge of capacitors

BASIC FACTS

- The time constant (usual symbol Greek tau, τ) of a charging or discharging circuit = CR.
- The charge Q on a capacitor of capacitance C, when the circuit resistance is R and the time is t, is given by

  $Q = Q_0 e^{-\frac{t}{CR}}$ for discharging, where Q_0 is the initial charge

  $Q = Q_0 (1 - e^{-\frac{t}{CR}})$ for charging, where Q_0 is the final charge

QUESTIONS

1 Show that the time constant $\tau (= CR)$ has the unit seconds when using SI units.

2 A 47 μF capacitor is connected to a 12 V supply. It is then disconnected from the supply and discharged through a 10 MΩ resistor. Calculate
(a) the time constant for the discharging
(b) the charge on the capacitor when connected to the supply
(c) the charge remaining on the capacitor after five time constants
(d) the potential difference across the capacitor after five time constants.

3 A 100 μF capacitor in a photographer's flash-gun is being charged from a 6.0 V supply through a 20 kΩ resistor. Calculate
(a) the charge that will eventually be stored by the capacitor
(b) the energy stored by the capacitor before the flash is operated
(c) the time constant of the charging circuit
(d) the time the photographer must leave between taking flash photos, assuming the capacitor must be at least 99% fully charged before using the flash.

4 A 15 μF capacitor is first connected to a 6.0 V supply and then disconnected. It is then discharged through a 500 kΩ resistor. Calculate
(a) the initial charge on the capacitor
(b) the time constant for the circuit
(c) the initial potential difference across the resistor
(d) the initial discharging current
(e) the time it would take the capacitor to discharge if this initial current were maintained.
If you have worked accurately through this question, you should have found that the answers to (b) and (e) are the same. This is true in general, and you should now be able to prove that it is always true that the initial discharging (or charging) current would completely discharge (or charge) a capacitor in one time constant.

5 Calculate the time it takes for a 100 μF capacitor to discharge to half its original charge when it is being discharged through a 20 kΩ resistor. (This figure corresponds to the half-life of the capacitor's discharge circuit. Half-life is not a term much used when dealing with capacitors, although the mathematically similar problems in radioactivity are frequently handled this way. Note that the time constant for a capacitor discharge circuit is not equal to the half-life and nor is it similar to the decay constant for radioactive decay.)

Charge and discharge of capacitors

6 An uncharged 0.010 μF capacitor is connected through a 2.2 MΩ resistor to a 9.0 V supply. Calculate

(a) the final charge on the capacitor

(b) the charge 18 ms after the charging starts

(c) the p.d. across the capacitor at this time

(d) the p.d. across the resistor at this time

(e) the current at this time.

7 A 4.7 pF capacitor in a computer needs to be charged from a 6.0 V supply to at least 5.0 V in a thousandth of a millionth of a second (10^{-9} s). What is the maximum resistance that can be used in the charging circuit?

Magnetic effect of an electric current

BACKGROUND INFORMATION

MAGNETIC FIELD

As with all fields, the definition of magnetic field strength is determined in terms of a force. In the case of magnetic field strength, the force used is the force exerted on a unit length of wire carrying unit current. The formal definition of magnetic field strength is:

Magnetic field strength (B) at a point is defined as the magnetic force per unit current in a wire at the point per unit length of the wire, when the wire is at right angles to the field. Magnetic field strength is also called magnetic flux density.

$$\textit{magnetic field strength} = \frac{\textit{magnetic force}}{\textit{current} \times \textit{length of wire}}$$

or, in symbols,

$$B = \frac{F}{Il} \quad \text{giving} \quad F = BIl$$

The unit of B, the N A^{-1} m^{-1}, is called the tesla (T).

If the wire is at any angle θ to the field, the equation becomes $F = Il \sin \theta$.

If the current exists as a result of there being a charge q moving with velocity v, then $F = BIl = Bql/t = Bqv$. Again, if the wire is at any angle θ to the field, the equation becomes $F = Bqv \sin \theta$.

The direction of these forces is given by the left-hand rule, in which the thumb, representing the direction of the force, the first finger, representing the direction of the field, and the second finger, representing the direction of the current, are held so that they are all perpendicular to one another.

MAGNETIC FIELD STRENGTH NEAR CURRENT-CARRYING CONDUCTORS

The magnetic field strength B near any arrangement of current-carrying conductors depends directly on the electric currents. B also depends on the geometry of the arrangement. A constant of proportionality is required in all the equations, and this constant, which is given the symbol μ_0, is called the permeability of free space. The numerical value of μ_0 is $4\pi \times 10^{-7}$ N A^{-2}. This is an exact value as a result of the way the ampere is defined. As with electric fields, these equations are designed so that no π term will appear in formulae involving uniform fields, a 2π term will appear in formulae involving cylindrical symmetry, and a 4π term will appear in any formula involving spherical symmetry.

(a) Magnetic field strength near a long straight wire carrying a current I

$$B = \frac{\mu_0 I}{2\pi r}$$ where r is the distance from the wire.

(b) Magnetic field strength at the centre of a flat, circular coil carrrying a curent I

$$B = \frac{\mu_0 IN}{2r}$$ where r is the radius of the coil and N is the number of turns on the coil.

(c) Magnetic field strength inside a long solenoid carrying a current I

$$B = \mu_0 nI$$ where n is the number of turns per unit length of the solenoid.

Magnetic effect of an electric current

LESSON POINTERS

It is assumed here that the work on magnetic field given in Topic 1 of Section 2 has not been taught. If it was taught at the commencement of Section 2, some of this lesson can be done as a quick revision.

There is a need to check some GCSE work before the start of this lesson. Students need to be able to sketch the magnetic field pattern around: (a) a wire carrying a current (and be able to use some method of determining the field's direction, such as the right-hand grip rule, in which the wire is held in the right hand with the thumb pointing in the direction of the current and the fingers are then pointing around the wire in the direction of the field); (b) a coil carrying a current; (c) a solenoid; and (d) a bar magnet.

The term magnetic field strength is used here as a descriptive term directly parallel to electric field strength. The term magnetic flux density is preferred by some teachers and examination boards. This is another situation in which students have to be aware of both terms so that they can use either. The symbol B is fortunately universally used.

The deduction of the equations given for magnetic field strength requires calculus and, even for these simple arrangements, the calculus is quite involved. A-level Physics syllabuses do not now require students to be able to deduce the equations or to memorise them, but they do expect students to be able to use them.

Magnetic effect of an electric current

ANSWERS TO WORKSHEET

1 1.6×10^{-5} T

2 (a) 0.145 T

(b)

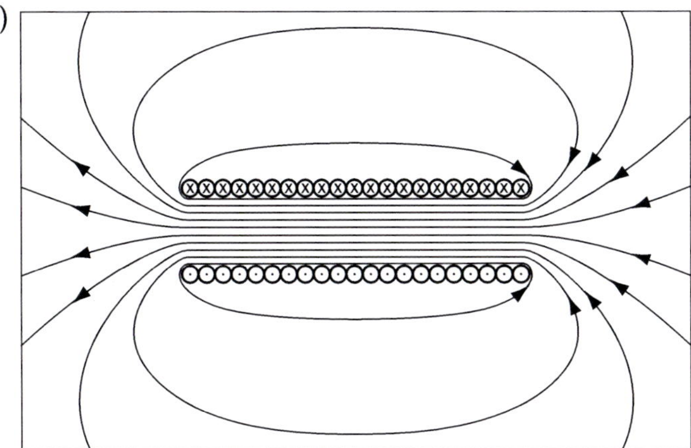

3 0.151 N

4 (a) 1.36×10^{-11} N

(b) 1.49×10^{19} m s^{-2}

5 (a) 4.3×10^{-4} T away from the observer

(b) 6.5×10^{-3} N m^{-1} away from A

(c) 6.5×10^{-3} N m^{-1} away from B

(d)

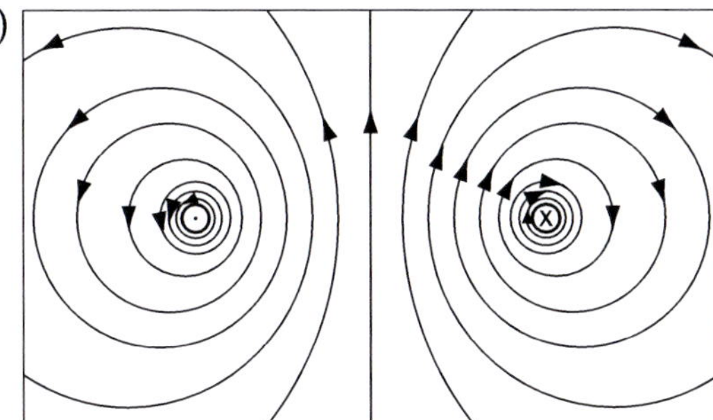

6 (a) Electric field exerts force on electron of 3.2×10^{-13} N upwards, so magnetic field must exert a force of the same magnitude downwards.

(b) 6.25×10^{6} m s^{-1}

(c) Force due to electric field is constant but force due to magnetic field is proportional to speed. A slow electron will therefore accelerate downwards, since the magnetic force is less. A fast electron will accelerate upwards.

Magnetic effect of an electric current

BASIC FACTS

- Force F on a wire of length l, carrying a current I in a direction at right angles to a magnetic field of magnetic field strength BI, is given by $F = BIl$. This is the equation that is used to define magnetic field strength. Magnetic field strength is also called the magnetic flux density.
- Force on a charge q, moving with velocity v at right angles to the field, is given by $F = Bqv$.
- These two equations become $F = BIl \sin\theta$ and $F = Bqv \sin\theta$, respectively, when the angle between the field and the current is θ.
- The direction of the force is given by the left-hand rule, in which the thumb represents the direction of the force (motion), the first finger represents the direction of the magnetic field, and the second finger represents the current.
- Magnetic field strength near a long straight wire carrying a current I

  $B = \dfrac{\mu_0 I}{2\pi r}$ where r is the distance from the wire.
- Magnetic field strength at the centre of a flat circular coil carrrying a curent I

  $B = \dfrac{\mu_0 IN}{2r}$ where r is the radius of the coil and N is the number of turns on the coil.
- Magnetic field strength inside a long solenoid carrying a current I

  $B = \mu_0 nI$ where n is the number of turns per unit length of the solenoid.

QUESTIONS

1 Calculate the magnetic field strength (magnetic flux density) at a distance of 0.15 m from a long wire carrying a current of 12 A.

2 (a) Calculate the magnetic field strength (magnetic flux density) inside a solenoid which is 0.20 m long and has 10 000 turns of wire. The solenoid current is 2.3 A and it can be considered as a long solenoid.

(b) The field inside such a solenoid is very uniform away from its ends, both along the axis and across its area of cross section. Perhaps surprisingly, the value of the magnetic field strength does not depend on the radius of the solenoid, but of course a greater length of wire will have current through it if the radius of the solenoid is large. Draw a diagram showing the magnetic field both inside and outside the solenoid.

3 A piece of wire in an electric motor is 300 mm long and carries an electric current of 7.4 A. The wire is at right angles to a magnetic field of field strength 0.068 T. Calculate the force exerted on the wire.

4 An electron (charge -1.60×10^{-19} C) travels with velocity of 3.7×10^{7} m s^{-1} at right angles to a magnetic field of field strength 2.3 T. Calculate

(a) the force exerted by the field on the electron

(b) the acceleration of the electron.

Magnetic effect of an electric current

5 Two long parallel wires X and Y are placed a distance of 0.0060 m apart. Wire X carries a current of 13.0 A upwards and wire Y a current of 15.0 A downwards as shown in the diagram.

(a) Determine the field (magnitude and direction) that wire X causes at wire Y.

(b) Determine the force per unit length (magnitude and direction) that this field causes on Y.

(c) Determine the force per unit length (magnitude and direction) that Y exerts on X.

(d) Draw a diagram showing the magnetic field in the space around the wires and show the direction of the force on both X and Y.

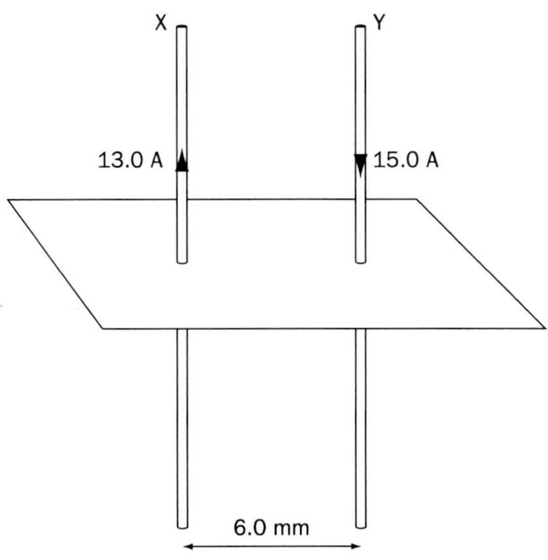

6 An electron is injected into a region in which there is an electric field of field strength 2.00×10^6 Vm^{-1} and a magnetic field of field strength 0.32 T. The two fields are arranged in the way shown in the diagram, and it is found that the electron travels right through the region with no deflection.

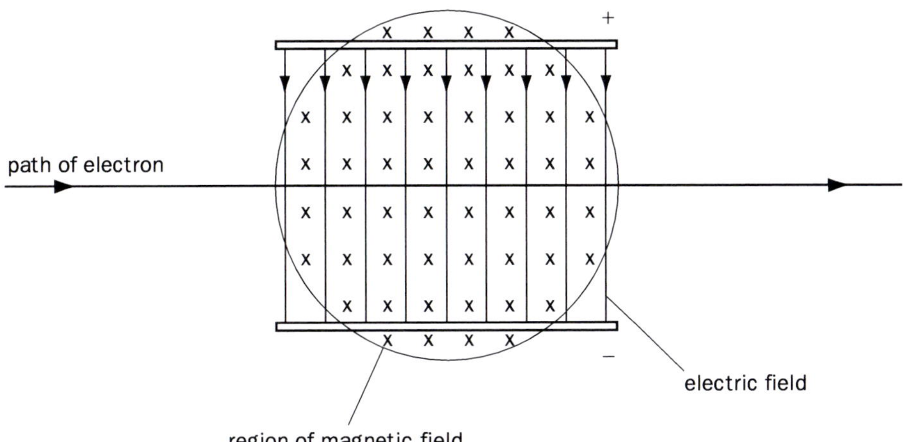

(a) Explain how it is possible for there to be no deflection.

(b) Calculate the speed of the electron through the region.

(c) Explain qualitatively what will happen to slower and faster electrons when injected into the region in the same direction as the electron in (a).

Electromagnetic induction

BACKGROUND INFORMATION

MAGNETIC FLUX AND MAGNETIC FLUX LINKAGE

Magnetic flux Φ is defined by the equation

magnetic flux = magnetic flux density × area

In symbols this becomes

$$\Phi = BA$$

The area here is the area at right angles to the magnetic flux density.

The additional term flux linkage may be needed for a coil. The flux linkage for a coil is simply the product of the flux through the coil and the number of turns N on the coil:

flux linkage for a coil = ΦN

FARADAY'S LAW OF ELECTROMAGNETIC INDUCTION

Once Oersted had discovered the magnetic effect of an electric current there was considerable scientific activity in the early part of the nineteenth century to make practical devices, such as electromagnets and electric motors. It took about twelve years before the reverse process of using a magnetic field to produce an electric current was discovered by both Faraday and Henry. The effect, called electromagnetic induction, is of enormous practical significance since it is used in all large-scale production of electrical energy. The effect can easily be demonstrated in the laboratory. It is suggested that a build-up from the simplest example to practical devices takes place in the following way.

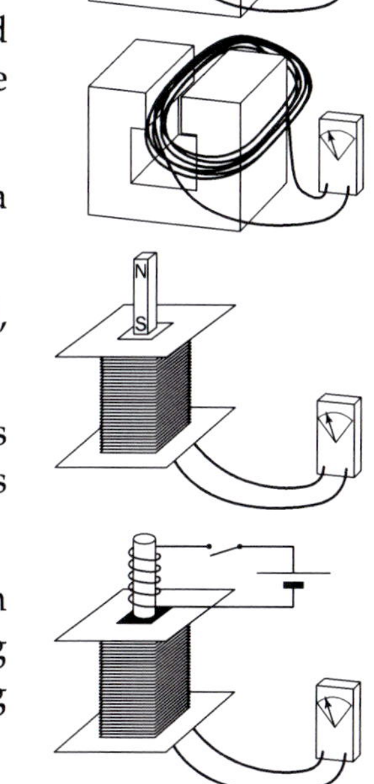

- Use a U-shaped magnet and a piece of wire connected to a very sensitive galvanometer. It can easily be seen that only when movement of the wire across flux takes place is an e.m.f. generated. The direction of the generated e.m.f. changes when the direction of movement across the field changes. The result is a dynamo effect and the right-hand dynamo rule can be mentioned.
- The effect can be increased by using more than one wire, i.e. by using a coil of wire, still with the sensitive galvanometer.
- This is more convenient with a bar magnet moving in and out of a coil, with a much larger number of turns.
- The magnet could stay still and the coil could move. The same effect is seen – movement is always necessary. Reversing the movement always reverses any current produced.
- The magnet could be an electromagnet. (Make sure it runs on a smooth d.c. supply – not a rectified a.c. supply.) Now you can show that switching the electromagnet on and off is just as effective (or more so) in changing the magnetic field as moving a magnet.

Electromagnetic induction

- The electromagnet could be connected to, say, a 6 V a.c. supply. The rapidly changing flux will produce rapidly changing e.m.f. across the coil. This can be done with a commercial demountable transformer.

All of the above demonstrations illustrate Faraday's law of electromagnetic induction, which states that *E*, the electromotive force (e.m.f.) produced, is proportional to the rate at which magnetic flux is cut. The theoretical proof, from the law of conservation of energy, that the constant of proportionality is one in SI units is not required at A-level. The law is best written in calculus notation, but for those students for whom this is off putting the use of Δ is worthwhile:

$$\textit{electromotive force (e.m.f.)} = \frac{\textit{flux cut}}{\textit{time}} = \frac{\Delta \Phi}{t}$$

$$E = -\frac{\mathrm{d}\Phi}{\mathrm{d}t}$$

LENZ' LAW

You will have noticed that a minus sign has been inserted in the above calculus equation. This sign indicates the direction in which any generated current will flow. Lenz' law states that the e.m.f. generated is always in such a direction as to cause any generated current to be in such a direction so as to oppose the motion which is producing it. This too is a consequence of the law of conservation of energy. The minus sign reminds students of this law.

Students should do some of the questions at this point, particularly those which involve graphical work on flux, rate of change of flux and induced e.m.f.

At this stage reference could also be made back to the demonstrations in the last section, in particular to numbers 3 and 4. When the south pole of the magnet is pushed down into the coil, any current in the coil will itself make the coil an electromagnet. The direction of this current must cause a south pole at the top of the coil in order to oppose the introduction of the magnet. When the magnet is extracted from the coil, the current in the coil will reverse to make the coil have a north pole at its top, in order to oppose the removal of the magnet. This is Lenz' law in action. Any current will always try to prevent motion.

There are many practical examples of this law. An extra safety braking system on many coaches and lorries uses the effect. An electrically conducting and moving part of the transmission system is placed near a large electromagnet. When the electromagnet is switched on, large currents are set up in the transmission system and they oppose the motion of the transmission system and hence brake the vehicle. These opposing forces are also important when considering the efficiency of transformers, electric motors and dynamos. It is a consequence of Lenz' law that motors and dynamos cannot be more than 100% efficient.

Electromagnetic induction

TEACHER NOTES

LESSON POINTERS

The advantage of using the term magnetic flux density rather than magnetic field strength becomes apparent at this stage. If 'density' is thought of as an area density rather than a volume density, it leads easily to the idea of flux as being the flux density multiplied by the area. Originally this was done in terms of concentration of field lines, so the field strength was the number of lines per unit area and the flux was the total number of lines. This old-fashioned approach to the topic may help some students to visualise the relationship between the quantities.

The term flux linkage is not used very much but can be useful when dealing with such practical devices as transformers. In order not to confuse the terms flux and flux linkage, it is suggested that the unit weber (Wb) is used only for flux and that 'Wb turns' is used for flux linkage, even though 'turns' will only be a number.

It is recommended that the term flux linkage is not used unless your syllabus specifically mentions it. It can easily be avoided and it is one more item with which students can get into a muddle.

The minus sign in the Faraday equation arises because of Lenz' law. Great care has to be taken when using the law in more advanced work to get the minus sign in the correct place because, of course, any term in an equation can have its sign reversed by transferring it to the other side of the equation.

Students often find the plotting of sketch graphs difficult. In questions such as numbers 5 and 6, they are plotting the differential of the first graph on the second graph. This is a useful and constructive exercise which is important in many Physics topics, so it should not be rushed.

9 Electromagnetic induction

ANSWERS TO WORKSHEET

1 (a) 4.0×10^{6} Wb
(b) 8.0×10^{-3} Wb turns

2 (a) 6.5×10^{-5} Wb
(b) 0.65 Wb turns

3 (a) 0.645 Wb
(b) 0.645 V
(c) Zero, there is no complete circuit

4 176

5

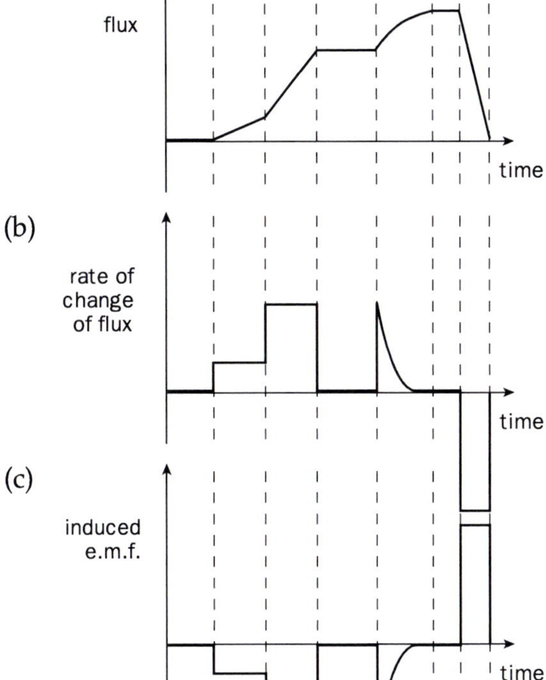

6

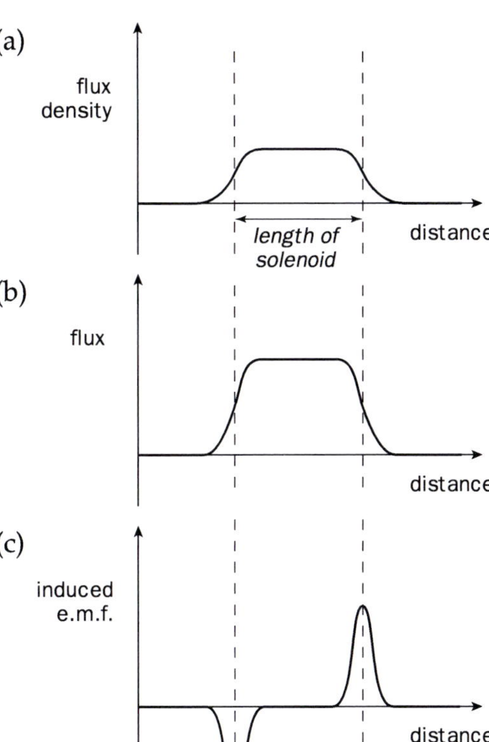

7 320 V

9 Electromagnetic induction

BASIC FACTS

- Magnetic flux Φ is defined by the equation

  magnetic flux = magnetic flux density × area

  In symbols this becomes

  $\Phi = BA$

  The area here is the area at right angles to the magnetic flux density.

  The additional term flux linkage may be needed for a coil. The flux linkage for a coil is simply the product of the flux through the coil and the number of turns N on the coil:

  flux linkage for a coil $= \Phi N$

- **Faraday's law of electromagnetic induction**

  The induced electromotive force (e.m.f.) is equal to the rate at which flux is cut. In equation form this becomes:

  $$\text{electromotive force (e.m.f.)} = \frac{\text{flux cut}}{\text{time}} = \frac{\Delta\Phi}{t}$$

  $$E = -\frac{d\Phi}{dt}$$

- **Lenz' law of electromagnetic induction**

  The e.m.f. generated by electromagnetic induction is always in such a direction so that any current it causes will oppose the motion which is producing it. Both of the laws of electromagnetic induction are a consequence of the law of conservation of energy. The minus sign in the above equation should remind you of Lenz' law. It shows that increasing flux being cut gives a negative e.m.f. and decreasing flux cut gives a positive e.m.f. Always be careful with the direction of induced e.m.f.

9 Electromagnetic induction

QUESTIONS

1 A solenoid with area of cross-section 0.00083 m^2 has 2000 turns of wire. The mean magnetic flux density through the solenoid is 0.0048 T. Calculate
(a) the flux through the solenoid
(b) the flux linkage through the solenoid.

2 Question 2(a) of Topic 8 read as follows:

'Calculate the magnetic field strength (magnetic flux density) inside a solenoid which is 0.20 m long and has 10 000 turns of wire. The solenoid current is 2.3 A and it can be considered as a long solenoid.'

The answer to that question is 0.145 T. Given an area of 4.5 cm^2 for that coil, calculate
(a) the flux through the solenoid
(b) the flux linkage through the solenoid.

3 The wing of an aircraft is 60 m long and the aircraft is travelling at 250 m s^{-1}. The vertical component of the Earth's magnetic field, which the wing is cutting, has the value 4.3×10^{-5} T. Calculate
(a) the flux cut per second by the wing
(b) the induced e.m.f. set up across the wing
(c) the current in the wing.

4 A wire in an electrical generator has length 3.2 m and is moving at a maximum speed of 8.6 m s^{-1} across a magnetic field of flux density 0.067 T. How many similar wires to this one will be necessary, if they are connected in series, for the induced e.m.f. across them to be 325 V?

5 (see following page)

6 A small coil is moved at a steady rate along the axis of a solenoid through which a current is passing. It starts beyond the solenoid, moves towards it, then through it and out at the other end. The plane of the coil is kept at right angles to the axis of the solenoid. Plot sketch graphs showing how the following vary with the distance moved:
(a) the magnetic flux density
(b) the flux through the coil
(c) the induced e.m.f.

7 An induction coil, which is used to cause the sparks in the sparking plugs of a car engine, has 4000 turns of wire and a mean area of cross-section 25 cm^2. A magnetic field of flux density 0.16 T passes through it. When the magnetic field is reduced steadily to zero in 0.0050 s, an e.m.f. is induced across the terminals of the coil. Calculate the value of this induced e.m.f.

Electromagnetic induction

5 The flux through a coil varies with time in the way shown in graph (a). Complete graphs (b) and (c) to show how the rate of change of flux and the induced e.m.f. respectively vary with time.

(a)

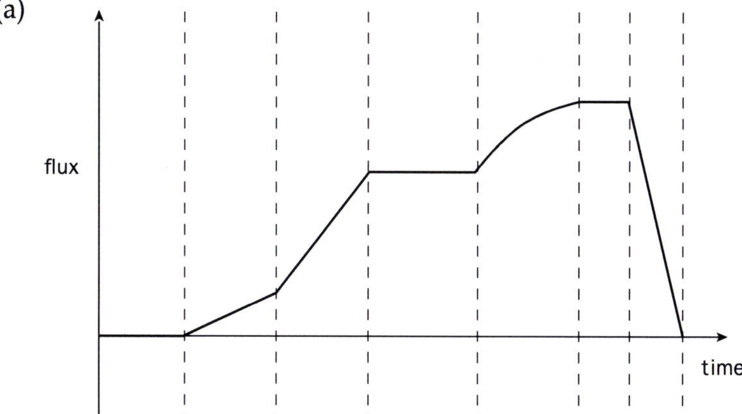

(b)

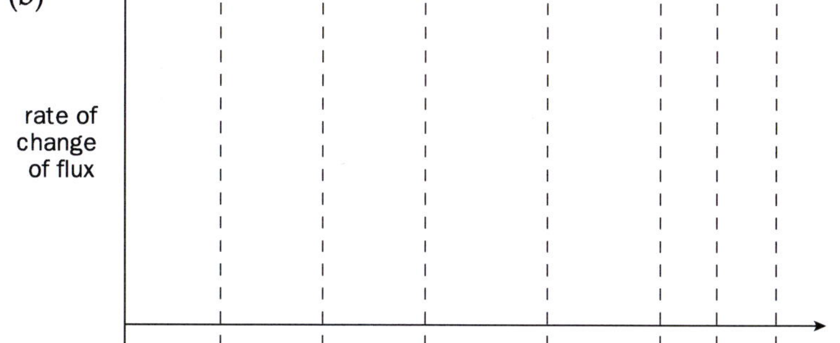

(c)

Alternating current (a.c.)

BACKGROUND INFORMATION

Alternating current (a.c.) is used for public electricity supplies throughout the world because of the need for transmission at high voltage. In order to obtain high voltage easily and with high efficiency it is necessary to use a transformer, and transformers will only work on a.c. This topic explains some of the terms associated with a.c. but it does not make any attempt to discuss features of a.c. theory associated with capacitors and inductors or with its importance in telecommunications. The term a.c. is often used in a useful descriptive way. In talking about this topic the expressions *a.c. current* and even *a.c. voltage* are heard, despite it being nonsensical to expand these terms to *alternating current current* and *alternating current voltage*.

POWER IN A RESISTOR WITH A.C. THROUGH IT

The laws of electromagnetic induction can be applied to a coil rotating in a uniform magnetic field and, as mentioned in Topic 9, the output e.m.f. from the terminals of the coil will be sinusoidal, as shown in Figure 1. If a resistor is attached to the terminals there will be a sinusoidal alternating current through the resistor, in phase with the alternating p.d. across the resistor.

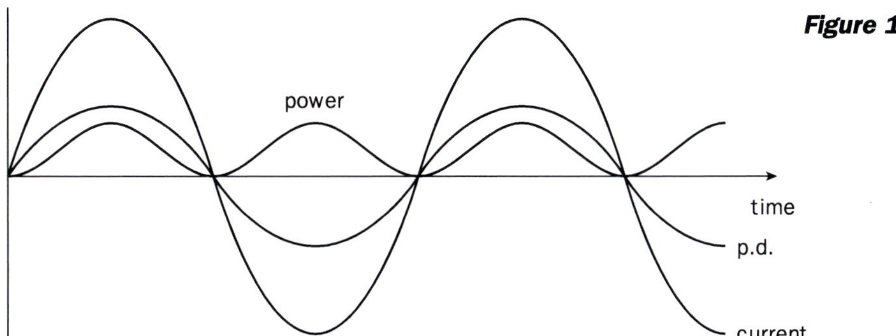

Figure 1

The way the power to the resistor changes is also shown on Figure 1. This is crucial to the understanding of r.m.s. values (see below). Two facts particularly should be emphasised.

1 When the p.d. is reversed the current is also reversed so the power is still positive.

2 The shape of the power curve is still sinusoidal so its average value is half its peak value.

This gives

peak power = peak current × peak e.m.f. or

$$\hat{P} = \hat{I} \times \hat{V}$$

therefore

$$\text{average power} = \tfrac{1}{2}\ \text{peak power} = \tfrac{1}{2}\hat{I}\hat{V} = \tfrac{1}{2}\hat{I}^2 R = \frac{1}{2}\frac{\hat{V}^2}{R}$$

The average power is the term usually required. A 100 W light bulb has an average power of 100 W although the instantaneous power varies from 0 W to 200 W 100 times per second if the mains frequency is 50 Hz.

10 Alternating current (a.c.)

TEACHER NOTES

It is clearly inconvenient in a.c. to use (average) power = $\frac{1}{2}I^2R$, and in d.c. to use power = I^2R. To overcome this problem the *root mean square current* $I_{\text{r.m.s.}}$ is used. It is preferable for students to think of this as a d.c. current. It is the d.c. current which has the same heating effect as the a.c. current. This gives

$$power = \frac{1}{2}\hat{I}^2R = I_{\text{r.m.s.}}{}^2R$$

and so

$$\frac{1}{2}\hat{I}^2 = I_{\text{r.m.s.}}{}^2$$

or

$$0.707\,\hat{I} = I_{\text{r.m.s.}}$$

A similar deduction for V gives $0.707\ V = V_{\text{r.m.s.}}$

Using r.m.s. values enables most a.c. calculations to be done in exactly the same way as those for d.c. The last question on the worksheet illustrates why high voltage is necessary for large-scale electrical transmission.

ANSWERS TO WORKSHEET

1 325 V

2 200 W, 0 W: 100 times, i.e. twice per cycle

3 (a) 5.30 A, 3.75 A
(b) 2250 W, 1125 W

4 (a) 12 400 V. This is very inefficient because of the 12 400 V supplied by the power station only 400 V are supplied to the factory, an efficiency of only 3.2%. The system is totally impractical.
(b) 40 120 V. This is very efficient. The corresponding efficiency is 99.7%.

Alternating current (a.c.)

BASIC FACTS

- Alternating current (a.c.) is used for public electricity supplies throughout the world because of the need for transmission at high voltage.
- In order to obtain high voltage easily and with high efficiency it is necessary to use a transformer, and transformers will only work on a.c.
- The term *a.c.* is often used in a useful descriptive way. In talking about this topic the expressions *a.c. current* and even *a.c. voltage* are heard, despite it being nonsensical to expand these terms to *alternating current current* and *alternating current voltage.*

Power in a resistor with a.c. through it

- Figure 1 shows the alternating current through a resistor when an alternating p.d. is applied across it.

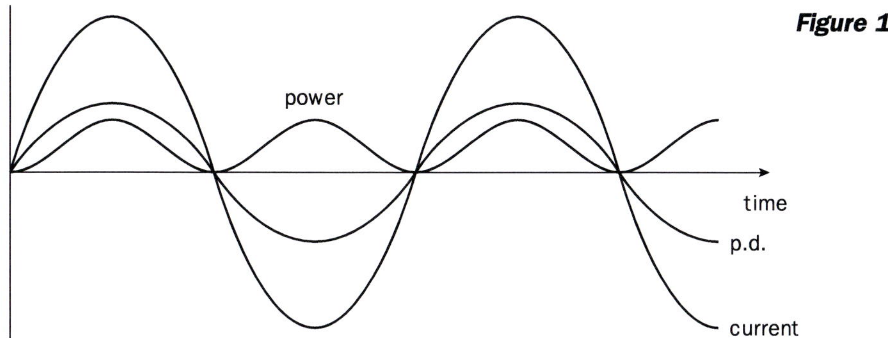

Figure 1

- The power graph can be obtained using the expression $P = VI$. When the p.d. is reversed the current is reversed too so the power is still positive. The power curve is still a sine wave shape so its average value is half its peak value.

  This gives

  peak power = peak current × peak e.m.f. or

  $\hat{P} = \hat{I} \times \hat{V}$

  therefore

  average power = $\frac{1}{2}$ peak power =

  $\frac{1}{2}\hat{I}\hat{V} = \frac{1}{2}\hat{I}^2 R = \frac{1}{2}\frac{\hat{V}^2}{R}$

- A term *root mean square current* is used. It is the d.c. current which has the same heating effect as the a.c. current. This gives

  power =

  $\frac{1}{2}\hat{I}^2 R = I_{\text{r.m.s.}}{}^2 R$

Alternating current (a.c.)

and so

$$\frac{1}{2}\hat{I}^2 = I_{\text{r.m.s.}}{}^2$$

or $0.707\,\hat{I} = I_{\text{r.m.s.}}$

A similar deduction for V gives $0.707\,V = V_{\text{r.m.s.}}$

Using r.m.s. values enables most a.c. calculations to be done in exactly the same way as those for d.c.

QUESTIONS

1 Calculate the peak value of the mains supply, which in the UK is 230 $V_{\text{r.m.s.}}$.

2 What are the peak value and the minimum value of the power to a 100 W mains bulb? With mains frequency of 50 Hz, how many times per second does a bulb reach peak power?

3 A resistor of resistance 80 Ω has an alternating voltage of peak value 300 V across it. Calculate
(a) the peak and r.m.s. current through it
(b) the peak and r.m.s. power supplied to it.

4 A power station supplies 1.2 MW to a factory through cables that have a total resistance of 4.0 Ω, as shown in Figure 2.

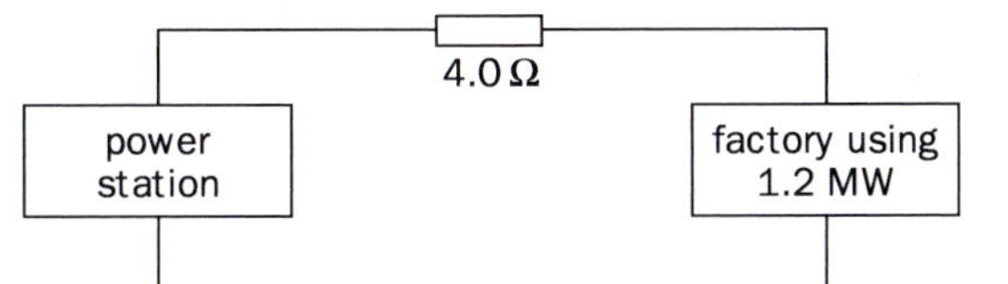

Figure 2

Calculate the power output required by the power station when the r.m.s. p.d. at the factory is
(a) 400 V
(b) 40 000 V

Comment on the practicability of your answers.

The transformer

BACKGROUND INFORMATION

Topics 10 and 11 show the principle and the need for a transformer. The laws of electromagnetic induction can be applied to explain why there is an output from the secondary coil of a transformer when there is an alternating current through the primary coil. Figure 1 shows the principle features of a transformer and its circuit symbol. In practice a transformer may be a few millimetres in size, handling a few microwatts of power, to a giant of many metres, handling many megawatts.

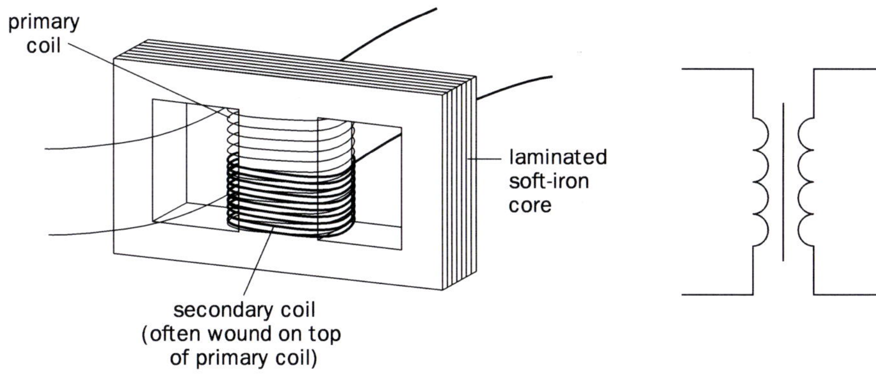

Figure 1

LESSON POINTERS

A demountable transformer is useful to show a transformer operating convincingly and students should be able to make their own transformers using pre-formed coils and C-cores. It is important to make the following points:

- the more turns there are on the secondary, the greater will be the secondary voltage;
- in the absence of anything given to the contrary, the electrical quantities given will be r.m.s. values;
- the power output of a transformer cannot be greater than the power input – so if the output voltage is increased then the output current must decrease;
- a step-up transformer has an output voltage greater than the input, whereas a step-down transformer has an output less than the input.
- turns ratio $= \dfrac{\text{number of turns on secondary}}{\text{number of turns on primary}}$

$$= \frac{\text{r.m.s. p.d. across secondary}}{\text{r.m.s. p.d. across primary}}$$

$$= \frac{\text{peak p.d. across secondary}}{\text{peak p.d. across primary}}$$

$$= \frac{\text{r.m.s. current through primary}}{\text{r.m.s. current through secondary}}$$

In symbols this becomes

$$r = \frac{n_s}{n_p} = \frac{V_s}{V_p} = \frac{I_p}{I_s}$$

The transformer

The assumption made to obtain the above equations is that the transformer is 100% efficient. In practice this is not a bad assumption as transformers often have efficiencies as high as 98%. Nevertheless, if a transformer is handling, say, 10 MW of power, the power loss is 200 kW and this will cause considerable heating of the transformer. Large transformers therefore require some mechanism to keep them cool. Transformers providing domestic supplies are usually cooled by incorporating oil-containing tubes in their construction that allow convection currents to circulate oil through and around the transformer.

The power losses in a transformer arise as a result of four factors.

1 Copper losses. The copper wires in the transformer have some resistance and therefore some heating takes place as a result of I^2R power loss. This effect can be reduced by using thicker copper wire but this has the effect of increasing the bulk of the transformer and making it more expensive. The modern trend is to make transformers smaller and allow them to become hotter, with consequent waste of power. This is possible now because of improved wire insulation. (All the wires in the transformer have to be insulated to prevent short circuiting.) Modern insulation, in motors as well as transformers, is a thin plastic film on the wire which does not melt even at high (200 °C) temperatures. The insulation can be made even thinner if the coils of wire in the transformer are wound in a pattern that keeps the p.d. between any two touching wires low, even in high-voltage transformers.

2 Eddy current losses. As a result of the rapidly changing magnetic flux in the core, eddy currents are induced in the core. These heat the core. The amount of eddy current heating is reduced by laminating the core.

3 Hysteresis losses. Each time the core is magnetised and demagnetised some energy is required. Since this is carried out 100 times per second in a mains transformer, power has to be supplied and this power results in heating of the core. This is minimised with modern materials which are very easy to magnetise and demagnetise.

4 Magnetic losses. Not all of the flux generated by the primary cuts the secondary. This can be minimised to a very small loss by good design and overlapping coils.

ANSWERS TO WORKSHEET

1 120 $V_{r.m.s.}$

2 $I_p = 7.3$ A, $I_s = 350$ A

3 (a) 30.7
(b) 769 turns
(c) 78 mA
(d) 3.6 mA

4 (a) 500 A
(b) 2940 kW, 60 kW
(c) 119 000 $V_{r.m.s.}$, 24.8 $A_{r.m.s.}$. Assume that the output voltage and current fall by the same fraction, i.e. to 0.99 of the value they would have had for a 100% efficient transformer.

The transformer

5 The following points should be included:

- the circuit allows the transmission to be made at high voltage
- therefore less current through the connecting cables
- less heating effect in the cables
- much more efficient transmission
- even with less than 100% efficient transformers
- a.c. has to be used rather than d.c.

11 The transformer

BASIC FACTS

- A diagram of a transformer is shown in Figure 1 together with its circuit symbol. A step-up transformer has an output voltage greater than the input whereas a step-down transformer has an output less than the input. Note that a transformer must always have at least four connections to it, two to the primary coil and two to the secondary coil. Transformers work only on alternating current.

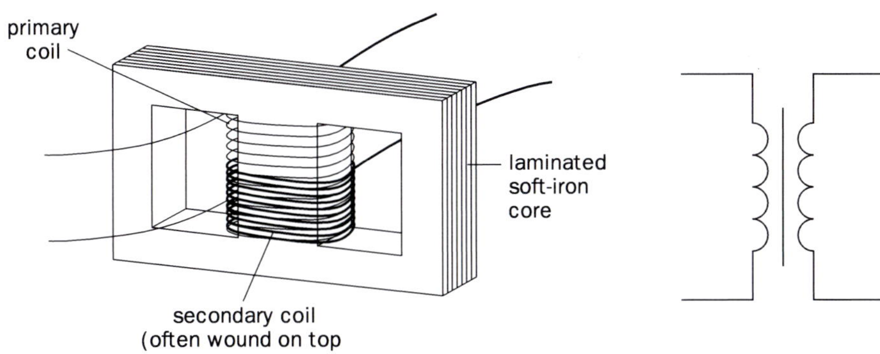

Figure 1

$$\text{Turns ratio} = \frac{\text{number of turns on secondary}}{\text{number of turns on primary}}$$

$$= \frac{\text{r.m.s. p.d. across secondary}}{\text{r.m.s. p.d. across primary}}$$

$$= \frac{\text{peak p.d. across secondary}}{\text{peak p.d. across primary}}$$

$$= \frac{\text{r.m.s. current through primary}}{\text{r.m.s. current through secondary}}$$

or $$r = \frac{n_s}{n_p} = \frac{V_s}{V_p} = \frac{I_p}{I_s}$$

- The assumption made in order to obtain the above equations is that the transformer is 100% efficient. In practice this is not a bad assumption as transformers often have efficiencies as high as 98%.

- The power losses in a transformer arise as a result of four factors.
  1. Copper losses. The copper wires in the transformer have some resistance and therefore some heating takes place as a result of I^2R power loss.
  2. Eddy current losses. As a result of the rapidly changing magnetic flux in the core, eddy currents are induced. These heat the core.
  3. Hysteresis losses. Each time the core is magnetised and demagnetised some energy is required.
  4. Magnetic losses. Not all of the flux generated by the primary cuts the secondary.

The transformer

QUESTIONS

Assume that all transformers referred to are 100% efficient unless it is stated otherwise.

1 A step-up transformer has 300 turns on its primary coil and 6000 turns on its secondary coil. Calculate the output voltage when the input is 6.0 $V_{r.m.s.}$.

2 A transformer for a street of houses has an input voltage of 11 000 $V_{r.m.s.}$ and an output of 230 $V_{r.m.s.}$. At a particular time it is supplying 80 kW to the houses. Calculate the primary current and the secondary current.

3 A transformer in a TV set provides a peak voltage of 10 000 V using a mains input at 230 $V_{r.m.s.}$. It has 25 turns of wire on its primary coil and provides 18 W of power. Calculate
(a) the turns ratio of the transformer
(b) the number of turns on the secondary
(c) the r.m.s. current to the primary
(d) the peak current from the secondary.

4 A step-up transformer with 3.0 MW input is 98% efficient. It has an input voltage of 6000 $V_{r.m.s.}$ and a turns ratio of 20. Calculate
(a) the input current
(b) the output power, and hence the power lost as heat in the transformer
(c) the output voltage and current.
Suggest how the transformer loses the heat generated within it.

5 Explain why it is preferable to use the circuit shown in the diagram for electrical power transmission rather than a circuit that does not contain any transformers.

SECTION
4
Waves

Simple harmonic motion

BACKGROUND INFORMATION

Oscillations have fascinated people since early times. The mathematics of oscillating pendulums or the imagined movement of molecules is complex and beyond the bounds of an A-level course. The model to be considered is a much-simplified version and, if students find the reasoning difficult, they should learn the equations and how to apply them and not worry too much about the proof.

For simple harmonic motion the acceleration is proportional to the displacement of an object from a fixed point, but in the opposite direction to the displacement. The constant of proportionality is written as ω^2 to ensure that it is a positive quantity, i.e.

$$acceleration = -\omega^2 x$$

Three examples follow which illustrate the ideas of simple harmonic motion and show how important it is to find ω.

EXAMPLE 1

A weight (mg) is suspended on a spring and, assuming that the spring obeys Hooke's law, we have

$$tension = k \times extension$$

where k is the spring constant in N m^{-1}. Looking at Figure 1, we can see $mg = k \times extension$. Let the extension equal r (this is for a static system). This gives

$$mg = kr$$

The suspended weight is now pulled down (i.e. work is done on the spring) a distance x. The weight is not in equilibrium and experiences an upward force F.

$$F - mg = -mass \times acceleration$$

from Newton's second law. The minus sign indicates that the force acts upwards in the direction of x decreasing (see Figure 2).

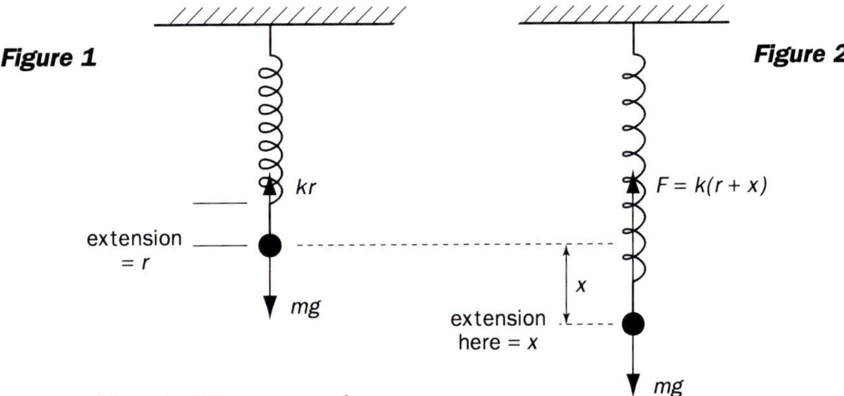

By Hooke's law, $F = k(r + x)$. We can say that

$$k(r + x) - mg = -m \times acceleration$$

Simple harmonic motion

We can substitute since $mg = kr$

$k(r + x) - kr = -m \times acceleration$

$kx = -m \times acceleration$

$acceleration = -\frac{kx}{m}$

Since the spring constant and mass of the object are both constant, the acceleration is proportional to the displacement but acts in the direction of displacement decreasing.

It is important to explain to students at this point that $\omega^2 = \frac{k}{m}$. Students can then work out ω in terms of k and m and then calculate f and T.

EXAMPLE 2

The second example which could be shown to a class is the simple pendulum (see Figure 3). The forces acting on the bob are the tension T in the string, and the weight mg of the bob.

Resolve the bob weight into a component perpendicular to the string (see Figure 3). The acceleration will act at right angles to the string.

$mg \sin\theta = -mass \times acceleration$

Once again the minus sign indicates that the force acts in the direction of θ decreasing. For small angles, $\sin\theta = x/l$.

$mg\frac{x}{l} = -mass \times acceleration$

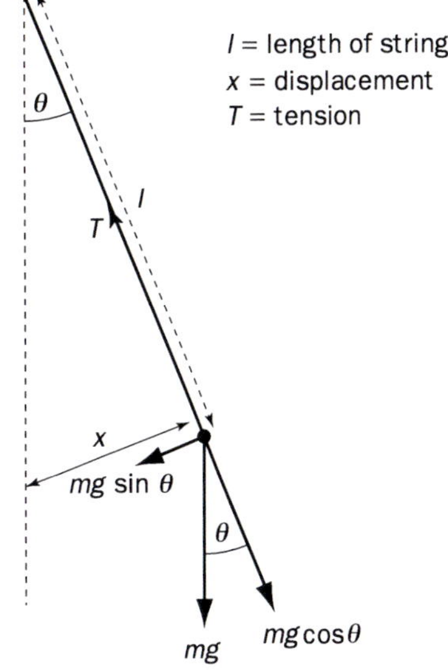

Figure 3

We can rewrite this as

$acceleration = -\frac{gx}{l}$

Once again, acceleration is proportional to displacement and in the opposite direction.

$\omega^2 = g/l$, since these are both constants for the system.

Students can work out T in terms of l and g. Discussion can take place and simple calculations can be done to show that this relationship holds for small angles (up to about 10°). Individual practical work can include pendulum experiments to investigate the T and l relationship. Oscillations with springs provide an opportunity to practise graph plotting and measurement skills.

Simple harmonic motion

EXAMPLE 3

The final example to demonstrate to a class involves circular motion. A turntable can be set up with a pencil supported vertically with tape or putty. The room is darkened and a light is shone on to the pencil. The turntable rotates and a shadow of the pencil's motion is observed on the wall. The oscillations are seen to go backwards and forwards from a central point.

The pencil moves at a constant angular velocity ω. Consider the pencil at position P and let O be the centre of the turntable, of radius r (see Figure 4).

$$\cos \theta = \frac{x}{r}$$

The pencil's shadow appears to move along the line AB. The displacement of the shadow from O is x, and the acceleration of the pencil is $\omega^2 r$ towards O. When this acceleration is resolved in the direction OB, we get

$$\textit{acceleration} = -\omega^2 r \cos \theta = -\omega^2 r \frac{x}{r}$$
$$= -\omega^2 x$$

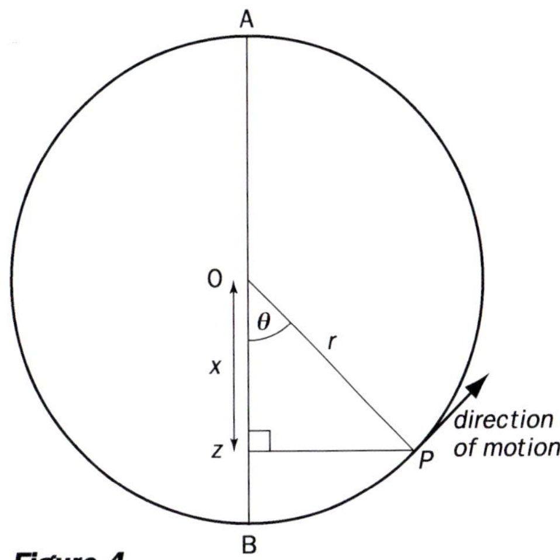

Figure 4

This again shows simple harmonic motion.

LESSON POINTERS

This is much more straightforward for non-mathematicians now that no formal proof is required. Students find it helpful at first if there are as many demonstrations of simple harmonic motion around the room as possible. It is worth setting up, for example,

- a weighted test-tube partially submerged in water
- a pendulum
- a mass on a spring
- a clamped steel ruler which can be twanged
- a ball-bearing on a concave lens
- a string clamped at both ends
- any musical instrument
- a loudspeaker connected to a sine wave oscillator of variable frequency.

Lots of practice using calculators for sine and cosine calculations will be needed for non-mathematicians.

Simple harmonic motion

If available, Sense and Control equipment is useful here to plot displacement, velocity and acceleration against time. A printout can be photocopied so that pupils have the same diagram in their notes for you to refer to in the lesson. Alternatively, for a displacement–time graph, a roll of wallpaper lining can be pulled at a constant rate under a sand-filled container, with a small hole in it. The container is swung to and fro, and a sine wave pattern is observed on the paper.

Once the students have the three graphs of the relationship between displacement–time, velocity–time and acceleration–time in front of them, they can look at phase relationships. It is a good idea to get them to plot the sine and cosine relationships given to emphasise the words amplitude and phase.

ANSWERS TO WORKSHEET

1 This question is mainly for non-mathematicians. They can be shown the quick way to plot the relationship, and it is to be hoped that they feel more comfortable in reading graphs in the future.

2 Once again this provides practice for non-mathematicians.
(a) (i) 1.0472 rad (ii) 2.3562 rad (iii) 0.1745 rad (iv) 4.7124 rad (v) 6.0912 rad
(b) (i) 57.30° (ii) 30.01° (iii) 45.02°

3 (i) 25.13 rad s^{-1} (ii) 0.754 m s^{-1} (iii) 0.653 m s^{-1}

4 (a) 1.269 s
(b) 0.4905 J
(c) kinetic energy at the maximum height 0 J, kinetic energy at the middle position 0.49 J
(d) 1.981 m s^{-1}
(e) 4.905 N

5 (a) 2.793 rad s^{-1}
(b) 0.0112 m s^{-1} or 11.17 mm s^{-1}
(c) 0.031 m s^{-1}

6 (a) 24.525 N m^{-1}
(b) 0.567 s
(c) 2.4525 N
(d) 1.4715 N
(e) 4.905 mJ
(f) 0.221 m s^{-1} or 221.5 mm s^{-1}

7 Plot T^2 against l. The gradient of the graph is $4\pi^2/g$.

Simple harmonic motion

BASIC FACTS

Oscillations have fascinated people since early times. The mathematics of oscillating pendulums or the imagined movement of molecules is complex and beyond the bounds of an Advanced level course. The model you have to consider is a much-simplified version and if you find the reasoning difficult, you should learn the equations and how to apply them and not worry too much about the proof.

Obvious to and fro motions are pendulums and pistons. You will find that examples will include rotating objects too. You will be able to see why these are used if you imagine a turntable rotating with a pencil attached to the table with blu-tack. If you shine a light on the pencil and observe its shadow, you will see as oscillating motion.

Terms used in simple harmonic motion

- *Frequency* is the number of oscillations per unit time. It is given the symbol f. The unit is the hertz (Hz). 1 hertz is equal to 1 cycle per second.
- *Time period* is the time for one complete oscillation. It is given the symbol T. The unit is the second.
- $T = 1/f$
- *Amplitude* is the maximum displacement in either direction from the central or mean position. It is a scalar, since the direction does not come into the definition. Here it is given the symbol x_0. The unit is the metre.
- If you plotted a graph of displacement–time for, say, a pendulum, the graph obtained would be a sine function. All simple harmonic motions can be considered to be of this form.
- A complete oscillation can be thought of to correspond to a complete circle. This repetitive pattern takes 2π radians to complete.
- The radian is a unit used to measure angles; 2π radians are equal to 360°. Half an oscillation is π radians (i.e. 180°).

  Look at the diagram right and see if you can relate it to the pendulum swinging.

  What is the difference between the broken curve and the continuous curve?

  For an oscillating body we can say that if θ is the angle moved through, and t is the time taken, then θ/t is the angular speed, ω.

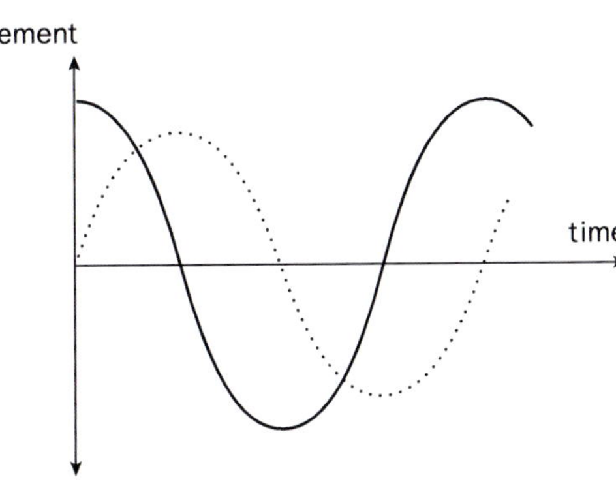

Simple harmonic motion

- *Angular frequency* (ω) is measured in radians and is related to T and f by

  $\omega = 2\pi/T = 2\pi f$

  Imagine a pendulum swinging to and fro. Think about where the maximum speed occurs. Is this where the maximum displacement is? The displacement is zero when the speed of the pendulum is at a maximum. The acceleration on the other hand is at a maximum when the displacement is at a maximum. The acceleration is always directed towards the centre (i.e. zero displacement).

- *Simple harmonic motion* is defined as motion in which the acceleration is proportional to the displacement of an object from a fixed point, but in the opposite direction to the displacement. For a sine wave, $x = x_0 \sin \omega t$, where x is the displacement at any time t, and x_0 is the amplitude.

  Imagine now that you are holding a weight on the end of a spring and letting the spring oscillate vertically up and down. The same ideas hold as for a pendulum. Maximum speed is again at the centre, where displacement is zero and so is acceleration. When the weight is at its lowest point, the acceleration is upwards towards the centre, and when the weight is at its highest point, the acceleration is once more acting in the opposite direction.

- Acceleration $\propto$ displacement.
  Let a = acceleration and x = displacement.

  $a \propto -x \quad a = -\text{constant } x$

  We will let the constant be ω^2

  $a = -\omega^2 x$

  This is the equation to learn and, for any system you are given, you should identify ω.

- Any oscillating system has energy. In the case of a pendulum the total energy of the system may be regarded as constant if we ignore friction. The energy changes occurring are from gravitational potential energy to kinetic energy and back again. At the mean position, we can consider all the energy to be in the form of kinetic energy, and, at the maximum displacement, we may consider that the energy is all gravitational potential energy. In between, there will be varying amounts of each but the sum will be constant.

- Finally, you can show that a motion is simple harmonic by proving that the acceleration is directly proportional to the displacement but acts in the opposite direction. This means that ω^2 can have any positive value, e.g. for a simple pendulum $\omega^2 = g/l$, where l is the length of the string, and g is the acceleration due to gravity; for a mass oscillating at the end of a spring, $\omega^2 = k/m$, where k is the spring constant, and m is the mass attached to spring.

- Other examples of simple harmonic motion can include oscillating loudspeaker cones, car suspension and pistons.

- When asked questions about simple harmonic motion, remember to think not only about the energy of the oscillating motion but also the energy of the system as a whole, e.g. when a mass is oscillating from a spring, the spring has some spring potential energy before it starts to oscillate, since the mass will have caused the spring to stretch.

1 Simple harmonic motion

Equations to learn

- Displacement: $x = x_0 \sin \omega t$
- Velocity: $v = x_0\, \omega \cos \omega t$ $\quad v = \omega\sqrt{(x_0^2 - x^2)}$
- Acceleration: $a = -x_0\omega^2 \sin \omega t$ $\quad a = -\omega^2 x$
- The kinetic energy of a body oscillating with simple harmonic motion is $\frac{1}{2}m\omega^2(x_0^2 - x^2)$.
- Simple pendulum: $T = 2\pi\sqrt{\frac{l}{g}}$ Mass on a spring: $T = 2\pi\sqrt{\frac{m}{k}}$
- The potential energy of a body oscillating with simple harmonic motion is $\frac{1}{2}m\omega^2 x^2$.

Look at the graph on the right to see the relationship between the kinetic energy and the potential energy of an oscillating body.

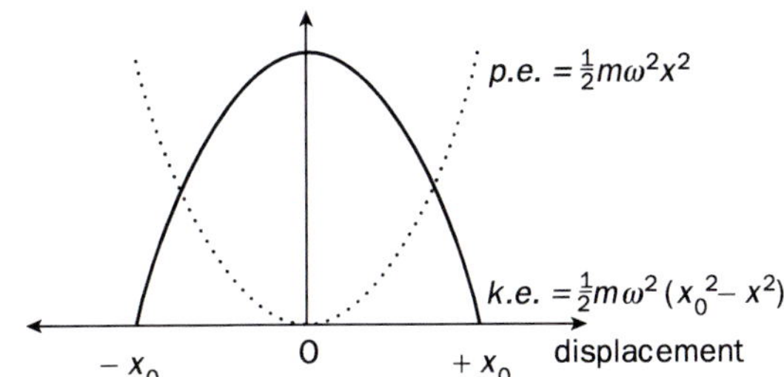

QUESTIONS

- Assume $g = 9.81$ m s^{-2}

1 Plot a graph of the following waveforms:
$x = x_0 \sin 2\pi ft$, where $f = 50$ Hz and $x_0 = 0.5$ m
$x = x_0 \sin 2\pi ft$, where $f = 100$ Hz and $x_0 = 1.0$ m
Now try to plot a cosine relationship for $x = x_0 \cos 2\pi ft$ for the same values of f and x_0.

2 (a) Convert these degrees into radians: 60°, 135°, 10°, 270°, 349°.
(b) Convert these radians into degrees: 1 rad, 0.5238 rad, 0.7857 rad.

3 A potato is hung from the lower end of a light vertical spring, whose upper end is fixed. The potato is pulled down and released and the time period of the oscillating system is measured. If the time period is 0.25 s and the amplitude of oscillation is 30 mm, calculate
(i) the angular frequency
(ii) the maximum speed
(iii) the speed when the displacement is 15 mm.

4 Half a coconut of mass 0.25 kg is suspended from a tree branch by a light piece of string of fixed length 0.40 m. A sudden gust of wind displaces the bob from its equilibrium position and the centre of mass of the coconut rises 20 cm above its rest position. Calculate
(a) the time period of the oscillations
(b) the potential energy as it reaches its maximum height
(c) the kinetic energy of the coconut at the maximum height and when it passes through the equilibrium position.
(d) the maximum speed of the coconut
(e) the maximum tension in the string.

Simple harmonic motion

5 A half-full bottle of lemonade is oscillating in the boot of a car and a wasp is floating on top of the liquid. If the amplitude of the oscillations is 4.0 mm and it takes 36.0 s to complete 16 complete oscillations, calculate

(a) the angular frequency of the system

(b) the maximum speed the wasp acquires

(c) the maximum acceleration.

6 A spring is suspended vertically from a clamp-stand and its length is measured. A mass of 0.2 kg is suspended from the lower end and the new length is noted. The mass is pulled down a further distance and then released from rest. The mass oscillates about the equilibrium position. How could you determine g using this apparatus? What other equipment would you require? State any precautions you would take to ensure accurate results and sketch the graph you would expect to obtain with labelled axes.

Natural length of vertical spring = 200 mm; extended length of spring in the rest position = 280 mm; length of spring when extended beyond rest position = 300 mm.

Use the information above to answer the following questions.

(a) Calculate the spring constant k.

(b) Find the time period of the oscillations.

(c) What is the maximum tension in the spring?

(d) What is the minimum tension in the spring?

(e) What is the maximum kinetic energy?

(f) What is the maximum speed?

7 You can use a simple pendulum to determine g. Sketch the graph you would plot and label the axes.

2 Free and forced vibrations

BACKGROUND INFORMATION

A simple way to introduce this topic is to have a pendulum set up in the laboratory. Figure 1 shows the additions that should be made.

Set the pendulum swinging, and ensure that all students observe the decrease in amplitude. You can then add a piece of card to the simple pendulum and set the pendulum swinging again. The pupils should observe the *damping effect*. Turn the card through 90° as shown in Figure 1, and let the students observe what happens. One position gives light damping and the other gives heavy damping.

Figure 1

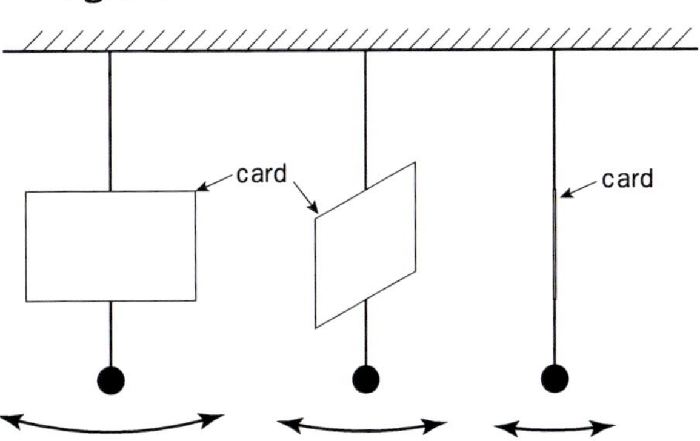

Figure 2

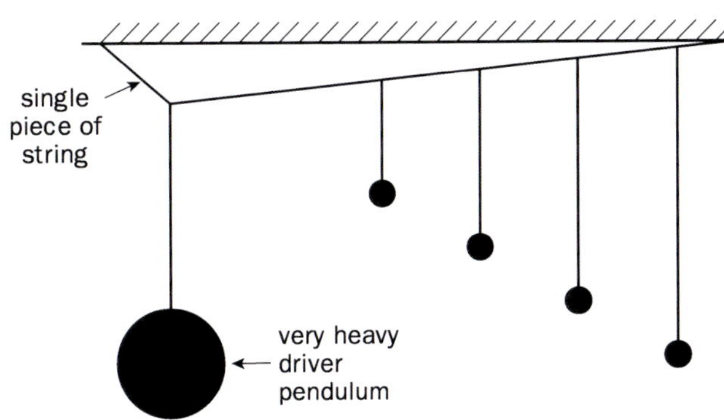

Then use the Barton pendulum set-up and set the driver pendulum swinging (see Figure 2).

The pupils should observe what happens to the other pendulums. All should see that the pendulums vibrate at the same frequency. Stop all motion and now set one of the other pendulums swinging. The class should make the same observation no matter which pendulum starts to swing first. The other pendulums are being forced to vibrate.

A third demonstration can be shown using two springs suspended vertically, with a mass between them and a vibrator at the bottom. The mass is made to oscillate and its natural frequency is found (see Figure 3).

The vibrator is attached to a signal generator. As the frequency is adjusted, resonance is observed and a graph can be plotted of amplitude of vibration against driver frequency. This should enable the students to see that the largest amplitude occurs at the natural frequency of the system. The mass can be damped in the same way that the pendulum was. Students should be able to see that the heavy damping alters the natural frequency of the system to a lower frequency.

Figure 3

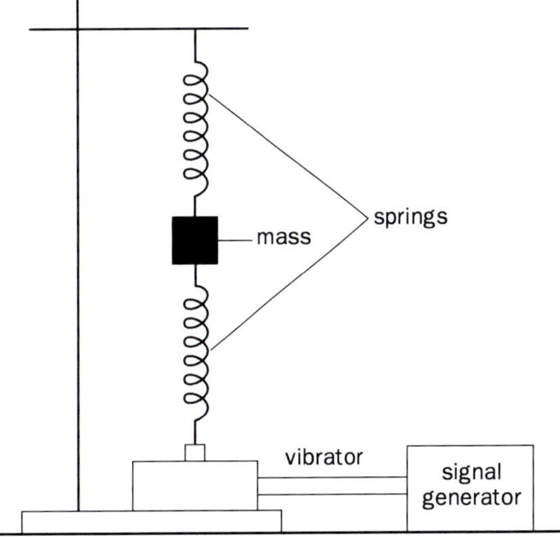

Free and forced vibrations

LESSON POINTERS

Discussion can include resonating wine glasses and opera singers shattering a glass by forcing it to vibrate at its natural frequency. Students tend to quote that Roman soldiers used to break step when crossing a bridge to prevent it resonating and collapsing (as recently as 1850, 200 soldiers were killed in Angers in France because a bridge they were marching on collapsed). Mention of the Tacoma Narrows bridge could be made here (this narrow bridge collapsed in 1939 after resonating at its natural frequency).

Microwave ovens make the water molecules in food resonate by applying a frequency equal to the natural frequency of the water molecules. Magnetic resonance is used by hospitals to scan the human body with, to date, no harmful effects. Aerials are constructed to resonate at the required signal frequency. Electric circuits also resonate at particular frequencies, and this is used to tune circuits.

ANSWERS TO WORKSHEET

Questions **1–7** involve the students' expressing their understanding in words. Some people find this hard to do, and it might be of benefit to the group as a whole if some responses were read out and the pupils decided who had expressed themselves most succinctly.

8 The time period increases.

Free and forced vibrations

BASIC FACTS

- Watch a child on a swing or washing blowing on a line – the amplitude of oscillation becomes smaller and smaller as the energy in the system gets less and less because of work being done to overcome resistive forces. This is known as natural damping. Unless energy is added to the system to overcome these forces, the oscillating object will eventually come to rest.

  Note: The time period has remained constant and independent of the energy in the system.

- Sometimes artificial damping is used in a system to reduce the time taken for the system to come to rest. The diagram below shows a naturally damped oscillation.

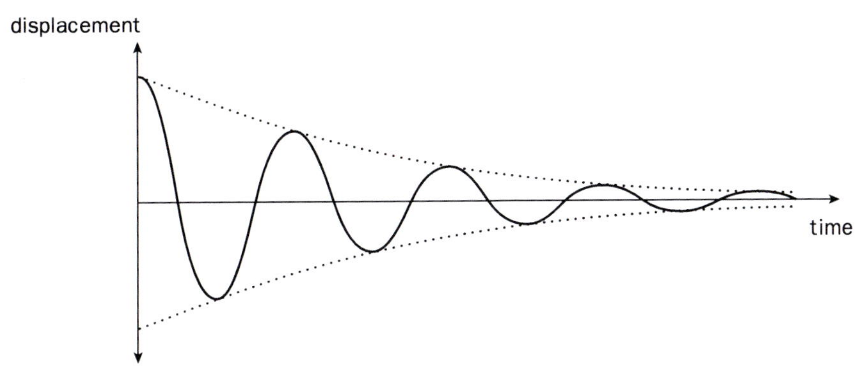

- Imagine you are playing the cymbals in an orchestra. You strike the cymbal and then you hold on to the cymbal to prevent it continuing to vibrate for too long. If you continued to hold on to the cymbal, it would not be free to vibrate when you next struck it.
- Shock absorbers are used to cut out unwanted vibration. Shock absorbers include your knees, which are useful when you land after jumping. Cars have shock absorbers to damp out the vibrations artificially, and they stop the ride becoming too bumpy.
- The suspension in a car is designed to give you as smooth a ride as possible. The springs ensure a less jarring ride and the oscillations in these are reduced by the shock absorbers.
- Damping can be light or heavy, depending on the particular system. A point is reached in damping a system beyond which the system stops oscillating. This is called critical damping. Ideally you require a car to have suspension which is just under critical damping. Think of other forms of transport that would benefit from damping.

Have you ever been in a car with a roof-rack fitted, and when the car reached a particular speed it started to make a sound?

Have you ever blown across a bottle that contained some liquid and heard a sound?

Have you observed divers at Olympic events bouncing up and down at the natural frequency of the board to give greater lift?

All of these are examples of resonance.

Free and forced vibrations

- Resonance occurs when the system is forced to oscillate at its natural frequency. This is simply seen by pushing a child on a swing. You apply the energy at the maximum amplitude.
- If a system had zero damping and you applied a forced frequency equal to the natural frequency, theoretically you should achieve infinite amplitude.
- You might have heard a car start to rattle at a particular speed; this is another example of resonance.
- If you plotted a graph of the amplitude of an oscillating system against the driver frequency, you would get a graph as shown below. Notice that the natural frequency of the system changes as the damping gets heavier.

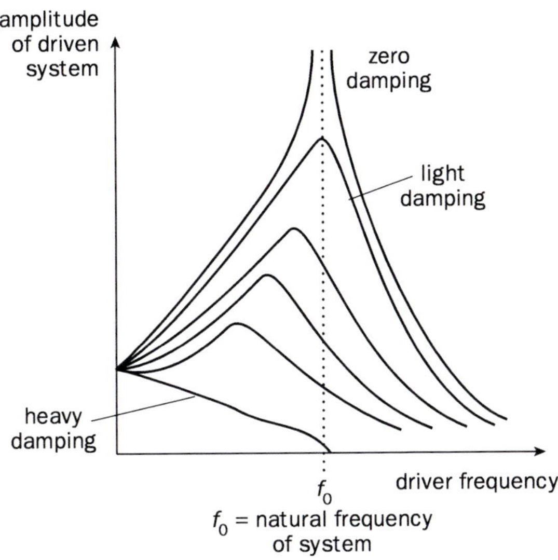

QUESTIONS

1 Sketch graphs to show the difference between damped and free oscillations. Label the *y* axis 'displacement' and the *x* axis 'time'.

2 What is meant by a free oscillation?

3 What is meant by a forced oscillation?

4 What is meant by resonance?

5 What does damping mean? How does the amount of damping effect resonance?

6 Describe an experiment that you could perform in the laboratory to show damping.

7 Sketch graphs on the same axes to show the difference between light, heavy and critical damping.

8 A girl is driving a light jeep and observes that, after hitting a bump, she oscillates with a period of about 1 s. She collects four friends for a night out. She travels over the same bump. Will the period of oscillation be more or less?

Wave properties

BACKGROUND INFORMATION

Energy can be transmitted from one point to another by waves. The wave energy causes a disturbance in the material or field, and this disturbance is passed on along the material or field. If it is a single pulse, it will gradually cause less and less disturbance until it ceases to exist (damping is covered in Topic 2). All waves, either *longitudinal* or *transverse* can be represented in one of two ways on a graph.

DISPLACEMENT–DISTANCE GRAPH

Imagine you could take a snapshot in time of a wave. At that instant some of the material or field would be experiencing their maximum disturbance, other parts would be undisturbed, and others would be somewhere in between. The graph of the displacement of the material or field from its equilibrium position at an instant is shown in Figure 1.

Figure 1

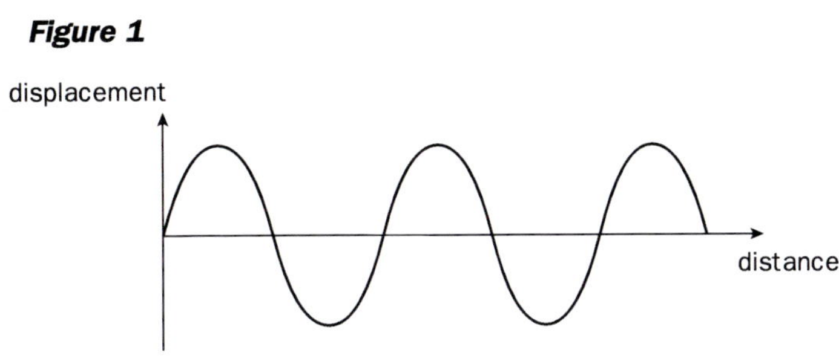

DISPLACEMENT–TIME GRAPH

If, on the other hand, you looked at just one particle or one point in a field, you would see how the oscillations from the equilibrium position varied with time.
See Figure 2.

Figure 2

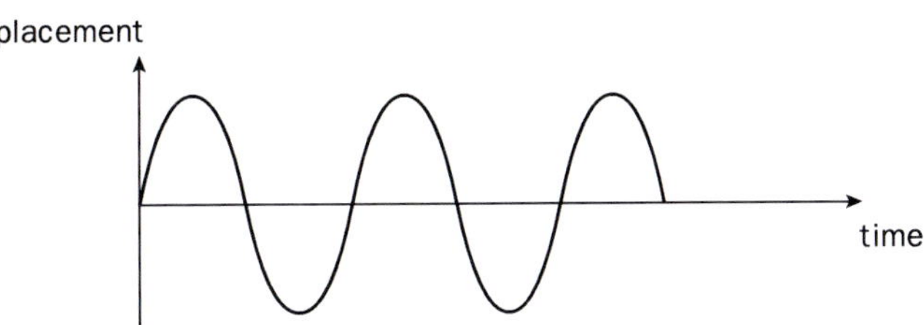

DEFINITIONS

Time period – time for one complete wave to pass a fixed point or the time for a particle to undergo a complete oscillation. The symbol is T and the unit is the second.

Amplitude – the maximum disturbance that a particle or field experiences as a wave passes. It has no symbol and the unit is the metre.

Wavelength – distance that one complete wave occupies. You measure from a particle to the next one that is in the same state of oscillation, e.g. from a crest to a crest. The symbol for wavelength is λ and the unit is the metre.

Frequency – the number of waves passing a given point in a second. The symbol is f and the unit is the hertz.

Wave properties

USEFUL RELATIONSHIPS

$f = 1/T$ $\qquad v = f \times \lambda$

where f is frequency in hertz, T is time period, v is wave velocity in m s^{-1}, λ is wavelength in metres.

PHASE AND PHASE DIFFERENCE

Look at Figure 3. In (a) the two waves have crests which coincide and they are said to be in phase. In (b) the crest of one wave coincides with the trough of another wave and they are said to be π radians out of phase, i.e. half a wavelength. In (c) one wave leads the other by a quarter of a wavelength, i.e. $\pi/2$ radians.

Figure 3

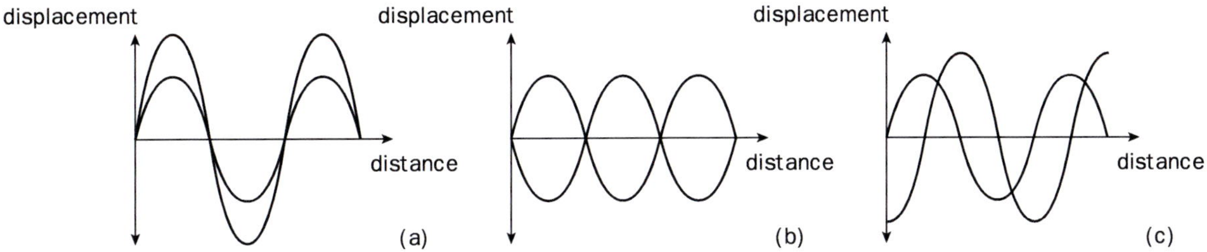

EQUATION OF A PROGRESSIVE WAVE

We looked at a sine wave graph when we did simple harmonic motion, we will now apply this same graph to a *progressive* or moving wave. (Think of a water wave in a ripple tank.) We can use the same equation as we used in that section, $x = x_0 \sin \omega t$, and x_0 is the amplitude if we think of looking at a disturbance–time graph. We can write this in a different form by substituting for ω. The equation now becomes

$$x = x_0 \sin 2\pi ft$$

If we had plotted the disturbance from equilibrium position for zero distance from the source, we could write our equation as

$$y = amplitude \times \sin 2\pi ft$$

If we call the amplitude a, we can find the disturbance y at any point x along the x axis. Any point on the wave will lag behind O by

$$\frac{xT}{\lambda}$$

Figure 4

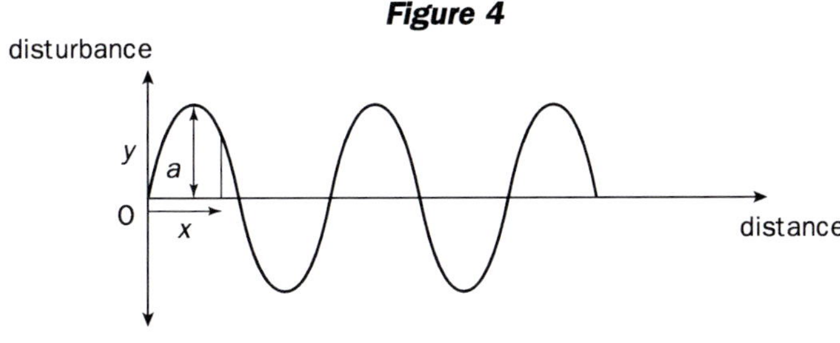

The equation becomes

$$y = a \sin 2\pi f\left(t - \frac{xT}{\lambda}\right)$$

Wave properties

Now substitute for $T = 1/f$ and the equation becomes

$$y = a \sin 2\pi\left(ft - \frac{x}{\lambda}\right).$$

So now we can find the disturbance at any time t, and at any distance x from the origin O! The phase angle between O and the point x in radians is given by $(2\pi x)/\lambda$.

SUPERPOSITION OF WAVES

Life gets even more complicated when we start to consider two waves meeting. We will look at the point at which they cross, because, once they have passed each other, they return to their original form.

- *The total disturbance at any point is the vector sum of the individual disturbances at that point.* (Vectors are used here because each wave has a speed and a direction.)

It is easy to see how this works if we take two waves in phase but with different amplitudes. The vector sum is simply the total of their amplitudes (see Figure 5).

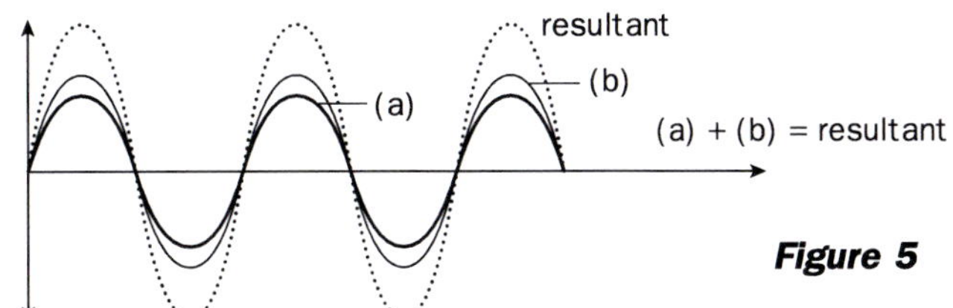

Figure 5

It becomes more difficult when the waves are not in phase (see Figure 6), but, with care, you could work out the disturbance for each wave at a particular time and add them together .

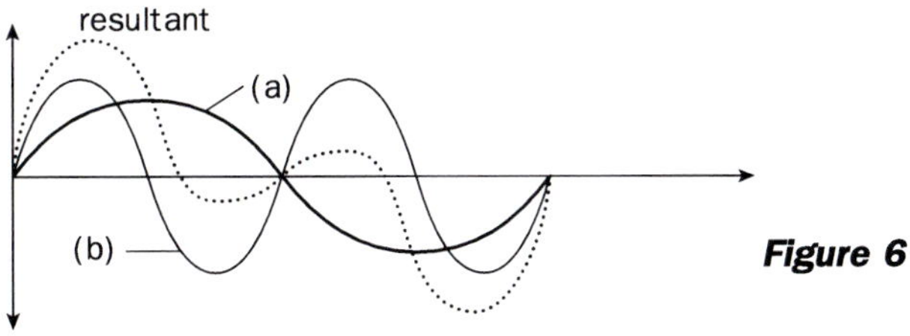

Figure 6

Examples of superposition include: interference, diffraction, stationary waves and beats.

INTENSITY OF WAVES

If the source of a wave is emitting the waves equally in all directions in three-dimensional space then the intensity I at a point a distance r from the source will decrease according to an inverse square law. Intensity is power per unit area and the surface area of the sphere through which the waves are travelling is $4\pi r^2$.

Therefore

$$I = \frac{P}{4\pi r^2}$$

Wave properties

LESSON POINTERS

The student sheets for this section are designed to be handed out in class or for homework, since most pupils will find the information a revision initially. The idea of phase difference can worry some people, but they should feel reassured when they have tried the simple questions. Students should feel confident with either displacement–time or displacement–distance graphs. However, the mathematics of the progressive wave equation, while interesting for keen pupils, could put weaker ones off. This part could either be omitted or gone through on the board with the teacher.

Superposition can be difficult for pupils to work out. I should aim for all to be able to cope with straightforward additions and let the keen pupil have a go at the more complicated vector addition.

ANSWERS TO WORKSHEET

1 Check that three complete oscillations have been drawn and that the period has been correctly marked.

2 Teacher's graph would give $x = x_0 \sin 2\pi ft$. Pupil's graph would give $x = x_0 \cos 2\pi ft$.

3 1.19×10^3 m

4 0.02 s

5 –0.019 m; the minus sign indicates downward displacement; phase angle is 1.6 rad.
Zero displacement and phase angle of 7π rad

6 (i) 3×10^{-3} m (ii) 82.8 Hz (iii) 4.0 m (iv) 331 m s^{-1} (v) 0.012 s (vi) yes

7 $y = 0.3 \sin(31.42t + 24.17x)$

8 1.62×10^{-6} W m^{-2}

Wave properties

BASIC FACTS

- Energy can be transmitted from one point to another by waves.
- The wave energy causes a disturbance in the material or field, and this disturbance is passed on along the material or field. If it is a single pulse, it will gradually cause less and less disturbance until it ceases to exist.
- All waves, either longitudinal or transverse can be represented in one of two ways on a graph. The displacement–distance graph:

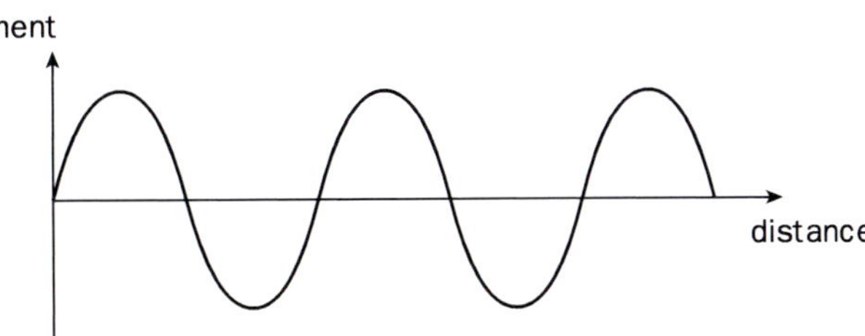

The displacement–time graph:

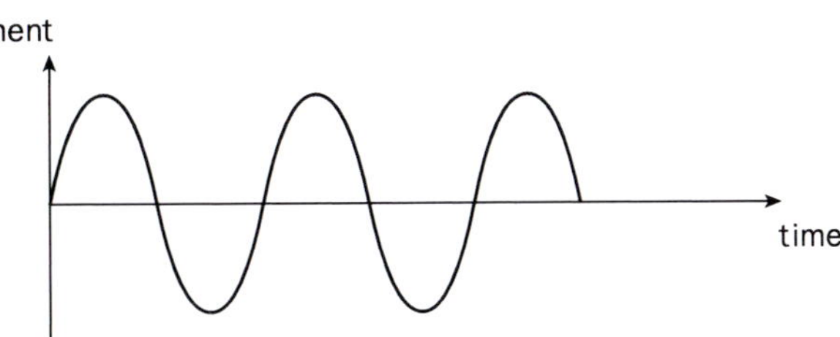

Terms you should know and use when answering wave questions

- *Time period* – time for one complete wave to pass a fixed point or the time for a particle to undergo a complete oscillation. The symbol is T and the unit is the second.
- *Amplitude* – the maximum disturbance that a particle or field experiences as a wave passes. It has no symbol and the unit is the metre.
- *Wavelength* – distance that one complete wave occupies. You measure from a particle to the next one that is in the same state of oscillation, e.g. from a crest to a crest. The symbol for wavelength is λ and the unit is the metre.
- *Frequency* – the number of waves passing a given point in a second. The symbol is f and the unit is the hertz.
- *Intensity* – the power of the wave passing through unit area. The symbol is I and the unit is W m^{-2}.

Wave properties

Useful relationships

- $f = 1/T$ $v = f \times \lambda$

  where f is frequency in Hz, T is time period in s, v is wave velocity in m s^{-1}, λ is wavelength in m.
- $I = \dfrac{P}{4\pi r^2}$

  where P is the power of, and r the distance from, the wave source.

Phase and phase difference

- Look at the sketches below. In (a) the two waves have crests which coincide and they are said to be in phase. In (b) the crest of one wave coincides with the trough of another wave and they are said to be π radians out of phase, i.e. half a wavelength. In (c) one wave leads the other by a quarter of a wavelength, i.e. π/2 radians.

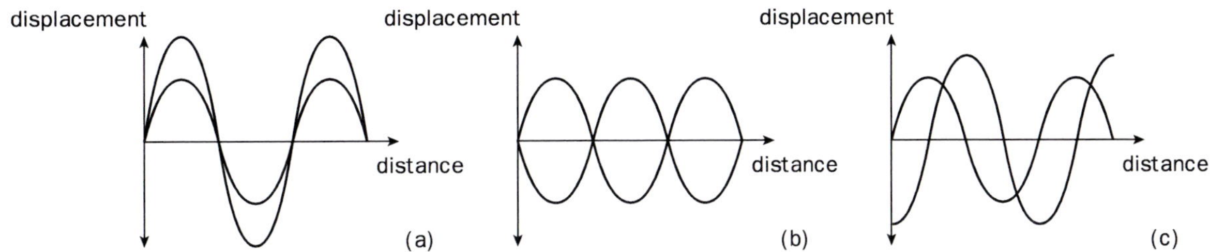

Equation of a progressive wave

- Apply the sine wave graph to a *progressive* or moving wave. (Think of a water wave in a ripple tank.) We can use the same equation as we used in the topic on simple harmonic motion. $x = x_0 \sin \omega t$, and x_0 is the amplitude if we think of looking at a disturbance/time graph.
- Substituting for ω. The equation now becomes $x = x_0 \sin 2\pi ft$. If we had plotted a disturbance from equilibrium position for zero distance from the source, we could write our equation as $y = amplitude \times \sin 2\pi ft$.

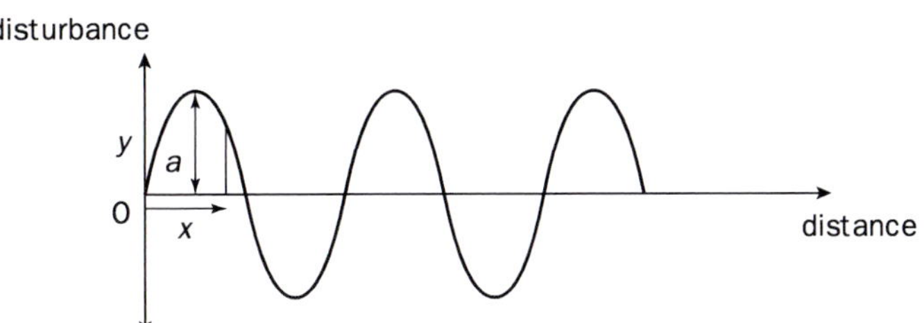

- If we call the amplitude a, we can find the disturbance y at any point x along the x axis. Any point on the wave will lag behind O by $(xT)/\lambda$.
- The equation becomes

  $$y = a \sin 2\pi f\left(t - \frac{xT}{\lambda}\right)$$

  Now substitute for $T = 1/f$ and the equation becomes

  $$y = a \sin 2\pi\left(ft - \frac{x}{\lambda}\right)$$

  So now you can find the disturbance at any time t, and at any distance x from the origin O!

  The phase angle between O and the point x in radians is given by $(2\pi x)/\lambda$.

Wave properties

Superposition of waves

- *The total disturbance at any point is the vector sum of the individual disturbances at that point.* (Vectors are used here because each wave has a speed and a direction.) It is easy to see how this works if we take two waves in phase but with different amplitudes. The vector sum is simply the total of their amplitudes (see diagram right).

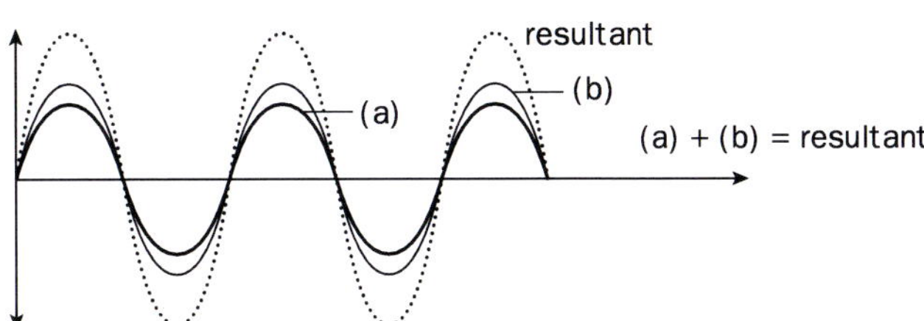

- When the waves are not in phase (see diagram right), work out the disturbance for each wave at a particular time and add them together.

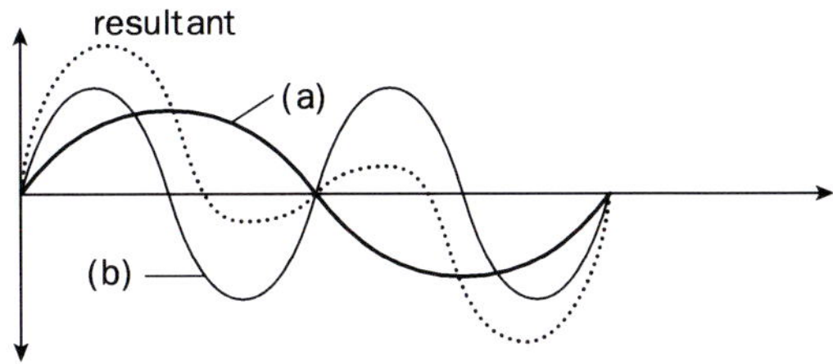

- Examples of superposition include: interference, diffraction, stationary waves and beats.

QUESTIONS

1 Draw a displacement–time graph for three complete oscillations of a wave with amplitude 3 cm and time period 2 s. Mark on your wave a crest and a trough, and show its period.

2 You are finding the time period for a pendulum. Your teacher tells you to start timing from the equilibrium position, but you time the swings from one of the maximum displacement positions. Sketch on the same axes the two graphs of displacement–time for the pendulum bob. Write an equation for each graph.

3 The speed of radio waves in air can be taken to be 3.0×10^8 m s^{-1}. If a radio station broadcasts on 252 kHz, what is the wavelength of the waves?

4 Electricity is transmitted at 50 Hz in Britain. What is the time period of this wave?

5 Water waves are viewed with a stroboscope. The wavelength is 10 cm and the amplitude is 2 cm. (Remember that $y = 0$ when $t = 0$.) Calculate the wave displacement at 8 cm and 35 cm. What are the phase angles at these points?

6 A sound wave can be written as $y = 3 \times 10^{-3} \sin (520t - 1.57x)$. Find (i) amplitude, (ii) frequency, (iii) wavelength, (iv) speed, (v) time period of the wave. (vi) Would the sound be audible?

7 Write an equation in terms of y, t and x for a water wave of amplitude 0.3 m, whose time period is 0.2 s and speed 1.3 m s^{-1}.

8 A television transmitter has an output of 100 kW and radiates equally in all directions. Calculate the intensity of the radio wave 70 km from the transmitter.

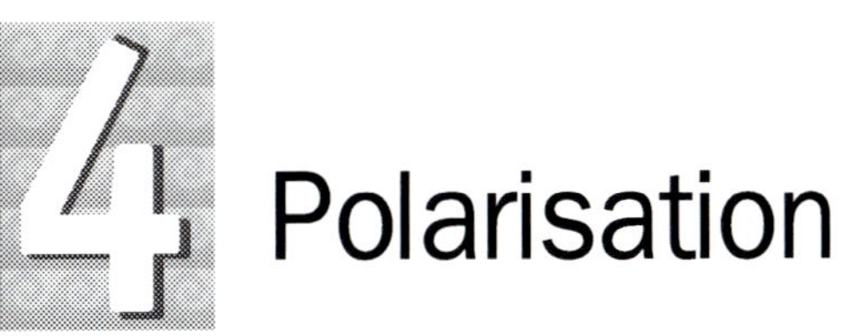

Polarisation

BACKGROUND INFORMATION

Waves can be divided into two categories, *longitudinal* or *transverse*. Longitudinal waves have vibrations that are in the same direction as the direction of travel (or propagation). Sound waves and compression waves are longitudinal. Transverse waves have vibrations at right angles to the direction of travel. Water waves, light waves, microwaves, radio waves are all transverse.

If we polarise a wave we reduce its intensity. Only transverse waves can be polarised, as you will see from the examples and the experiments that will be performed in the laboratory.

All electromagnetic waves have vibrating electric and magnetic fields. We will only consider electric fields here. Consider a light bulb emitting light. The varying electric fields can be resolved into two planes at right angles to the direction of travel. If you now pass the light through a polariser, the vibrations in one plane will not be transmitted (see Figure 1).

Figure 1

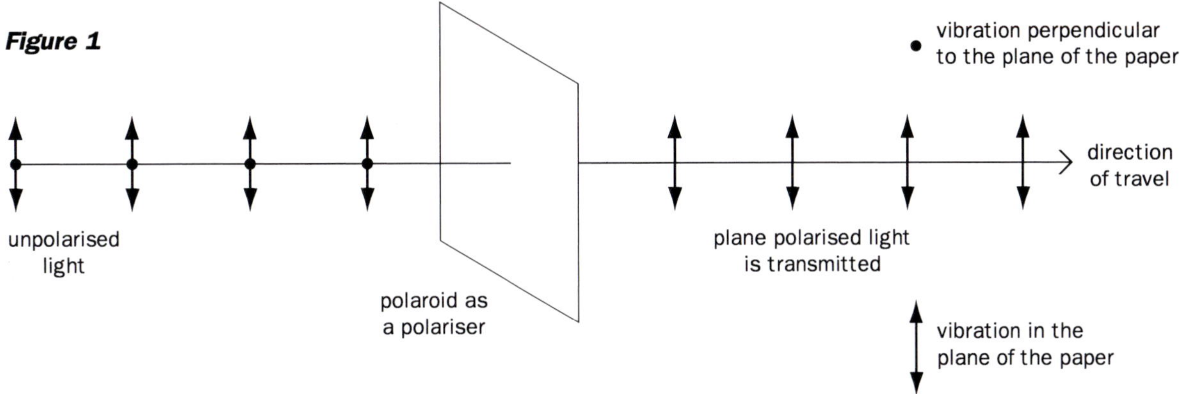

Polaroid film is a material that allows vibrations in one plane only. It contains crystals of quinine iodosulphate, which only transmit vibrations that are in a particular direction. Sunglasses reduce the intensity of the light falling on the eye by absorbing vibrations in one direction.

When light is reflected off a surface, it is partially plane polarised. However, you can shine a ray lamp at a glass block or a Perspex box filled with water and you can observe the reflected ray through a piece of Polaroid film. At one particular angle, the reflected ray completely disappears, showing that, at this angle, the ray was completely plane polarised. This occurs when the angle between the incident ray and the refracted ray is 90°. The angle of incidence that gives this plane polarisation is known as the Brewster angle and it is related to the refractive index of the material.

Polaroid film is used in stress analysis to pin-point weaknesses in structures. You can try this yourself if you have two pieces of Polaroid film between which you insert, for example, a Perspex set-square. If you press on the set-square and put it under stress, you observe interference fringes. Engineers make models of structures and use this method to look for weak points.

Polarisation

TEACHER NOTES

LESSON POINTERS

Most students will have heard of the name Polaroid in relation to sunglasses, and this makes an ideal starting point. The transverse nature of light will have been covered at GCSE level, but a slinky makes a good visual reminder of the difference between longitudinal and transverse waves.

Two clamp stands can be clamped together and a long piece of rubber tubing can be inserted between the gap. One student supports the clamp stands and another holds one end of the tubing. The teacher then sends transverse waves down the tubing and, depending on the direction of oscillation, the wave either passes through the gap or is stopped by it. This demonstration has two advantages in that it demonstrates Polaroid grating and the three-dimensional nature of an electromagnetic wave.

The commercially available microwave set-up can be used here, but the waves emitted are already polarised. Pieces of Polaroid film can be passed round, and students can rotate two pieces on top of each other to verify that all wave energy is absorbed by two pieces placed at right angles.

Strain viewers are available from suppliers, or students can use two pieces of Polaroid film and place Perspex or other transparent plastics between them and observe the interference fringes. Calculation of the Brewster angle can be carried out using an empty plastic box filled with water, a piece of Polaroid film and a ray lamp. Milk can then be added and the experiment repeated.

ANSWERS TO WORKSHEET

1 Place a metal grid in front of a loudspeaker, firstly in the horizontal and then in the vertical plane. A listener on the opposite side of the room will observe no difference in the sound intensity.

2 The Polaroid sunglasses absorb components of the vibrations of light that are in a particular direction. The intensity of light entering the eye is therefore reduced.

3

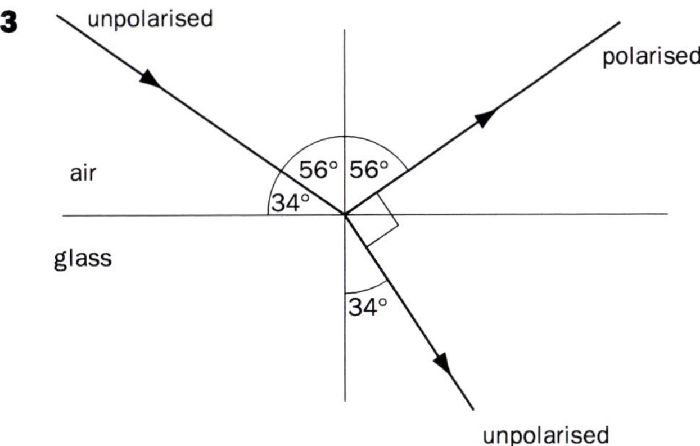

4 1.48

5 The reflected ray is the polarised ray, and this can be verified by observing the light through a piece of Polaroid film. As the Polaroid film is rotated, the ray will be completely absorbed by it at one point.

Polarisation

BASIC FACTS

- Waves can be divided into two categories, longitudinal or transverse.
- Longitudinal waves have vibrations, which are in the same direction as the direction of travel (or propagation). Sound waves and compression waves are longitudinal.
- Transverse waves have vibrations at right angles to the direction of travel. Water waves, light waves, microwaves, radio waves are all transverse.
- If we polarise a wave we reduce its intensity; only transverse waves can be polarised.
- All electromagnetic waves have vibrating electric and magnetic fields.
- Polaroid film is a material that allows vibrations in one plane only. It contains crystals of quinine iodosulphate, which only transmit vibrations that are in a particular direction. Sunglasses reduce the intensity of the light falling on the eye by absorbing vibrations in one direction.
- When light is reflected off a surface, it is partially plane polarised.
- The angle of incidence that gives this complete plane polarisation is known as the Brewster angle, and it is related to the refractive index of the material. This occurs when the angle between the incident ray and the refracted ray is 90°.
- Polaroid film is used in stress analysis to pin-point weaknesses in structures.

QUESTIONS

1 How could you demonstrate to a friend that sound waves cannot be polarised?

2 How do Polaroid sunglasses reduce the glare from the sunlight reflected off the sea?

3 Complete the diagram below to show the path of the refracted ray and the reflected ray, given that the angle between the reflected ray and the incident ray is 90°. Say which of the rays are polarised and which are unpolarised.

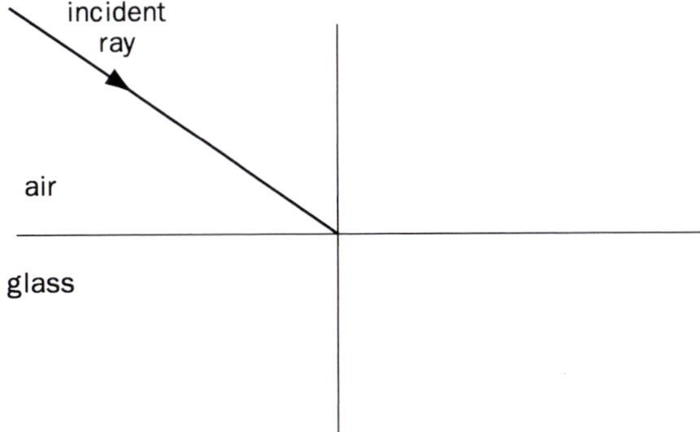

4 For the question above calculate the refractive index of the material.

5 How would you check which of the rays were polarised?

5 Diffraction

BACKGROUND INFORMATION

Diffraction is a phenomenon that is observed with all types of waves, be they transverse or longitudinal. When we have drawn ray diagrams or shown light rays being reflected from a mirror, we have used straight lines to represent the rays. The lines we have drawn are at right angles to the wavefronts. We are going to consider wavefronts and their motion in this section.

Waves spread round edges; if they did not you would have vertical edges to waves once they had travelled through a slit. The energy spills out around the gap. You can use a ripple tank to show this. The spilling out is more noticeable if the gap is approximately the same as the wavelength of the wave. Hence, you can hear conversations through half-open doorways because of the diffraction of sound. To observe diffraction in light, the gap must be very small, since the wavelength of light is about 5×10^{-7} m.

DIFFRACTION IN A RIPPLE TANK (SEE FIGURES 1 AND 2)

Diffraction is also observed when a wave encounters a barrier and the energy spills out around it and into areas that you would think of as shadow.

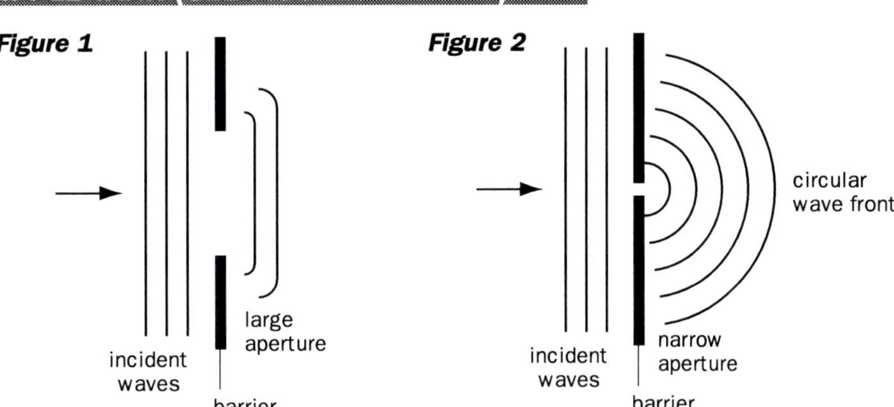

DEFINITIONS

Wavefront – a line showing the position of the crest of a wave.

Ray – a straight line drawn at right angles to the wavefront which shows the direction in which the ray is travelling.

Secondary wavelets – these can be constructed by considering that every point on a wavefront acts as a source of secondary waves. A time t later, these new waves will have travelled a distance equal to *wave speed* $\times$ t, and the new wavefront touches all these secondary wavelets (see Figures 3 and 4).

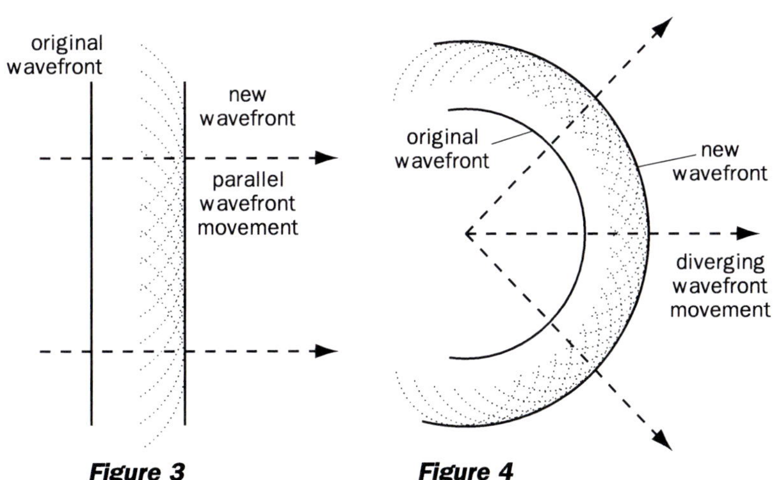

Figure 3 Figure 4

Diffraction

Superposition of waves – the total displacement due to waves travelling across a given point at a given time is the vector sum of the individual displacements produced by each wave. This is not as hard as it sounds: look at Figures 5 and 6 and you will see the outcome of two waves interacting.

Figure 5

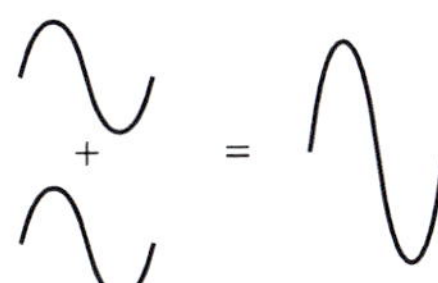

Figure 6

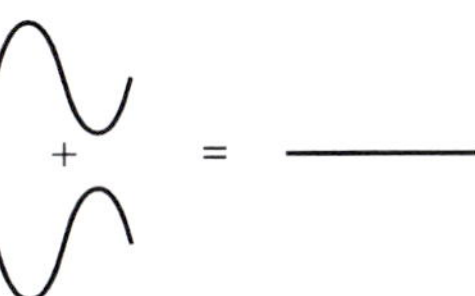

Imagine a plane wave approaching a narrow gap. The wave energy at the gap can be considered to be a source of secondary wavelets. The diffraction pattern observed is a result of the superposition of these secondary wavelets.

Diffraction is a phenomenon observed when a wave passes through a gap or passes round an obstacle. The spreading out of the wave energy is due to the superposition of the secondary wavelets from points along the same wavefront.

In everyday life, we encounter diffraction in sound waves. In the laboratory, we use specially produced slits or gratings because the wavelength of light is so small.

DIFFRACTION THROUGH A SLIT

Consider a beam of coherent light (one which has one frequency and a constant phase relationship and roughly the same amplitude) arriving at a gap of width a. We can consider that each point on a wavefront in the gap acts as a source of secondary wavelets. Consider just two of these (see Figure 7).

Figure 7

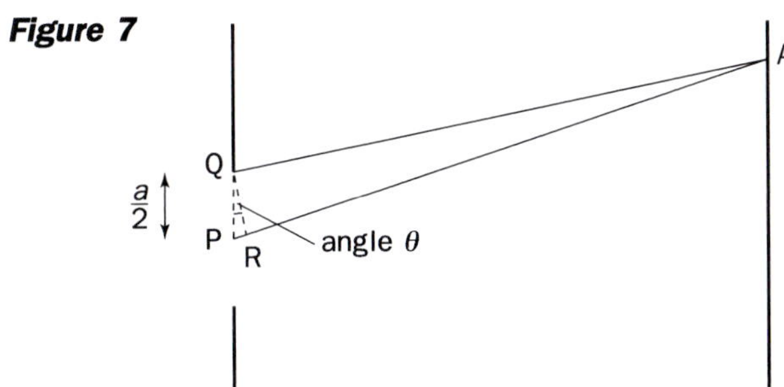

At point A, the rays will have travelled different distances, and there will be a path difference between them. If this is half a wavelength, the resultant displacement is zero. All other waves originating from the gap will have the same path difference and, therefore, nothing is seen at A.

$PR = \lambda/2$ $QP = a/2$ $\sin\theta = \lambda/a$

This gives the equation $a \sin\theta = \lambda$. This is the relationship for the first minimum. The second minimum is $a \sin\theta = 2\lambda$, etc. The bright spots or maxima would be halfway between the minima.

Diffraction

The intensity of the pattern decreases on each side of the central bright spot (see Figure 8). As *a* gets smaller, the diffraction pattern becomes more spread out.

Figure 8

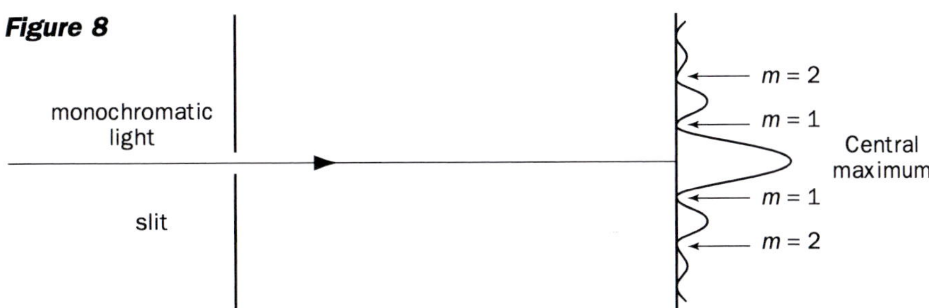

DIFFRACTION GRATING

We can then observe the patterns obtained from having a greater number of slits and these are manufactured to a high precision. The gratings usually have 300 or 600 lines per millimetre on them. Imagine a beam of parallel monochromatic light striking a grating, each gap will behave like a single slit. If you were to observe the emerging beam, you would see a bright central spot or maximum. As you move to either side of this spot you see a series of bright lines.

Figure 9

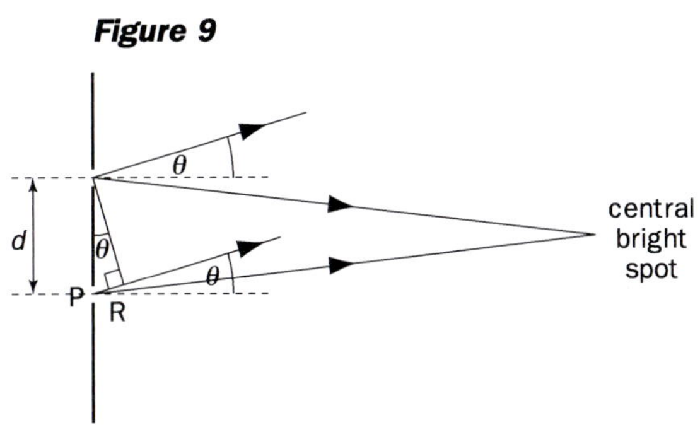

Let d equal the distance from the centre of one slit to the centre of an adjacent slit. Let θ be the angle through which the wave is diffracted. For a bright line to be observed, the path difference between the two waves must be a whole number of wavelengths, e.g. PR = $n\lambda$ (where n = 1, 2, 3, 4, …) (see Figure 9).

$$d \sin\theta = n\lambda$$

The number of bright lines seen depends on the wavelength of light used and the spacing between the slits. If white light is used, the central maximum is white, but the lines become bands of colour which can overlap. The shortest wavelength is seen nearest the centre.

Diffraction patterns have to be considered when designing cameras, microscopes and telescopes. They limit the size of the aperture and the size of the object that can be viewed with a microscope. Light cannot be used to observe atoms. Electron microscopes are used for this purpose.

LESSON POINTERS

This is another topic that is dealt with superficially at GCSE, but ripple tank demonstrations of the phenomena are useful for emphasising the importance of aperture width. It is important to explain secondary wavelets to pupils so that they can have a better understanding of interference and diffraction. Students who are interested might enjoy looking at reflection and refraction with this method too.

5 Diffraction

Students need to have the terms used in this section explained carefully to them, and they should be encouraged to use them in both written responses and oral work.

The laser is a useful tool to use with diffraction gratings in order to show diffracted spectra to the whole class at the same time, and to produce maxima, which are widely spaced and can be measured easily. The relationship between number of lines per millimetre and fringe spacing is clear to all the class. Talcum powder can be sprinkled on to the emerging beams from the grating, and the extent of the diffraction becomes clear.

ANSWERS TO WORKSHEET

1 (b) 1.67×10^{-6} m
(c) 2 orders
(d) central maximum is at 0° to the incident beam, first order is 22.9° to the incident beam and second order is 51.1° to the incident beam. The first and second orders appear each side of the central maximum.

2 First order = 10.4° second order = 21.1° third order = 32.7°
fourth order = 46.1° fifth order = 64.2°
Only five orders are seen with white light, 6.8°–13.0°.

3 509 nm, two orders are observed and, when dark blue light is used, more orders are seen, since it has a shorter wavelength.

4 17.7×10^{-3} m

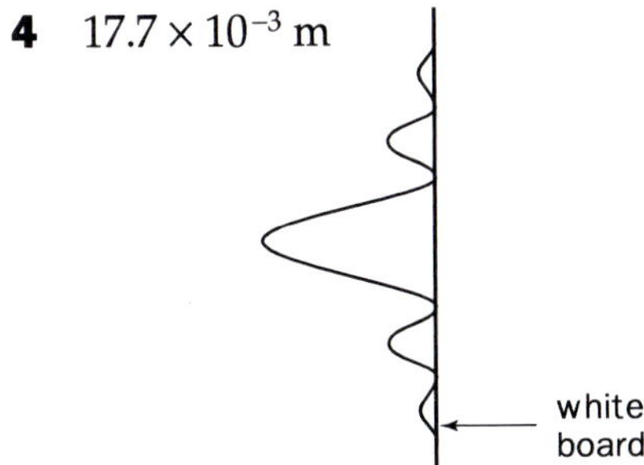

Diffraction

BASIC FACTS

- Diffraction is a phenomenon observed when a wave passes through a gap or passes round an obstacle. The spreading out of the wave energy is due to the superposition of the secondary wavelets from points along the same wavefront.

- *Wavefront* – a line showing the position of the crest of a wave.
- *Ray* – a straight line drawn at right angles to the wavefront; this shows the direction in which the ray is travelling.
- *Diffraction in a ripple tank* – diffraction is observed when a wave encounters a barrier, and the energy spills out around it and into areas that you would think of as shadow.

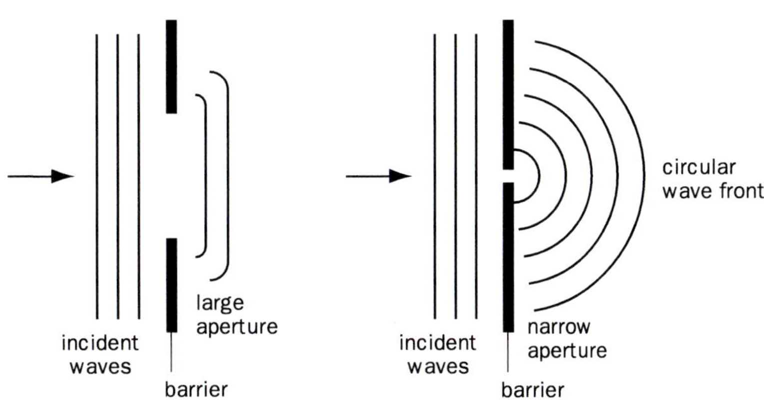

- *Secondary wavelets* – these can be constructed by considering that every point on a wavefront acts as a source of secondary waves. A time t later, these new waves will have travelled a distance equal to *wave speed* × t, and the new wavefront touches all these secondary wavelets (see diagrams).

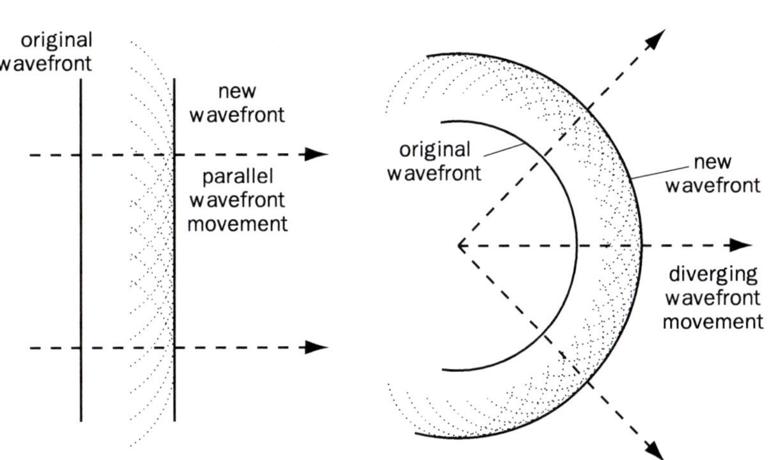

- *Superposition of waves* – the total displacement due to waves travelling across a given point at a given time is the vector sum of the individual displacements produced by each wave. This is not as hard as it sounds, look at the diagrams below, and you will see the outcome of two waves interacting.

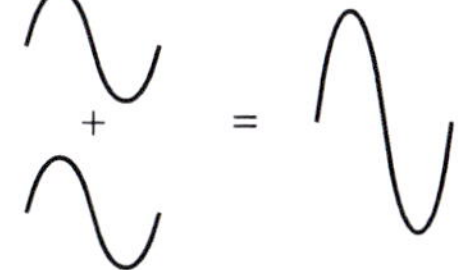

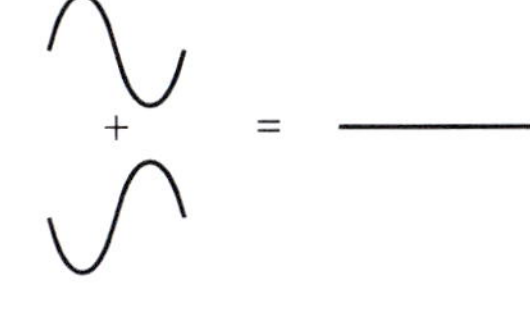

- *Diffraction through a slit* – minimum intensity is found at $a \sin \theta_m = m\lambda$, where λ is the wavelength, a is the slit width, and m is the order number of dark fringes from the centre.

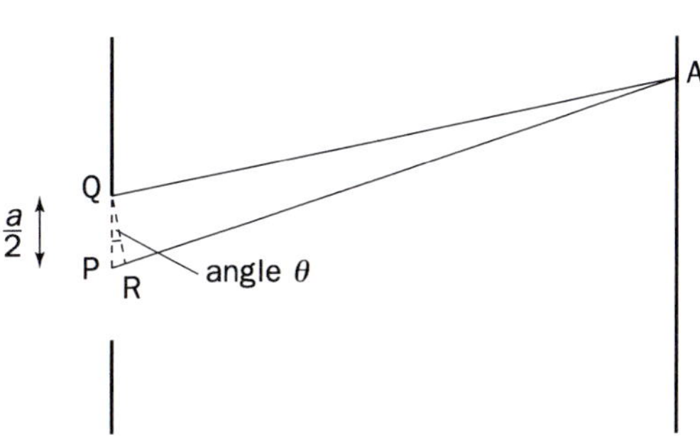

Diffraction

- *Diffraction grating* – maximum intensity is found at $d \sin \theta_n = n\lambda$, where λ is the wavelength, d is the slit spacing, and n is the order number of bright fringes from the centre.

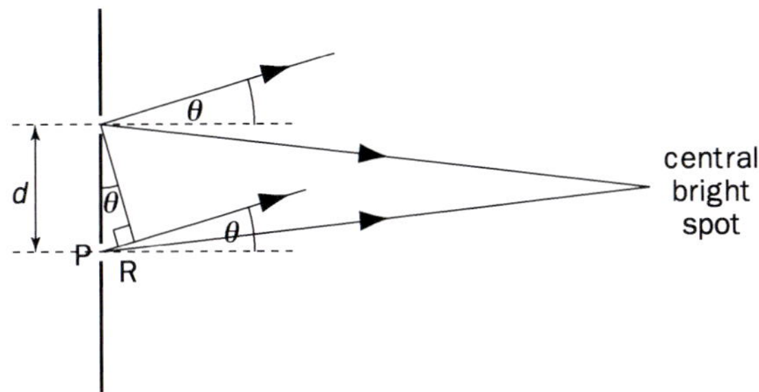

- Diffraction patterns have to be considered when designing cameras, microscopes and telescopes. They limit the size of the aperture and the size of the object that can be viewed with a microscope. Light cannot be used to observe atoms. Electron microscopes are used for this purpose.

QUESTIONS

1 A diffraction grating has 600 lines per millimetre. Monochromatic light of wavelength 650 nm is incident normally on it.
(a) What do the words 'incident normally' mean?
(b) What is the spacing d of the slits?
(c) How many diffracted beams would you observe?
(d) At what angle is each maximum formed?

2 A student shines a laser of wavelength 600 nm on to a grating, which has 300 lines per millimetre. Sketch what the student sees and label the angles between the maxima. She now replaces the laser with a source of white light (wavelength range 400–750 nm). How many orders would she see now? What would be the spread of the first order?

3 Monochromatic light falls on a grating, which has 500 lines per millimetre. The second-order line was observed at 30.6° to the central line. What is the wavelength of the light and how many orders are observed? If dark blue light had been used would more or fewer orders have been seen?

4 A teacher sets up a single-slit diffraction pattern to show his pupils. The sodium light used has a wavelength of 5.9×10^{-7} m, and it shines on to a slit of width 0.10 mm. The students observe the diffraction pattern on a white board 3.0 m away. Complete diagram (a) to show how the observed brightness varies across the board.

Calculate the distance between the centre of the central maximum and the first minimum. Use diagram (b) to help you, and make the assumption that for small angles $\theta = \sin \theta$.

Note: The width of the central maximum is twice that of the other maxima.

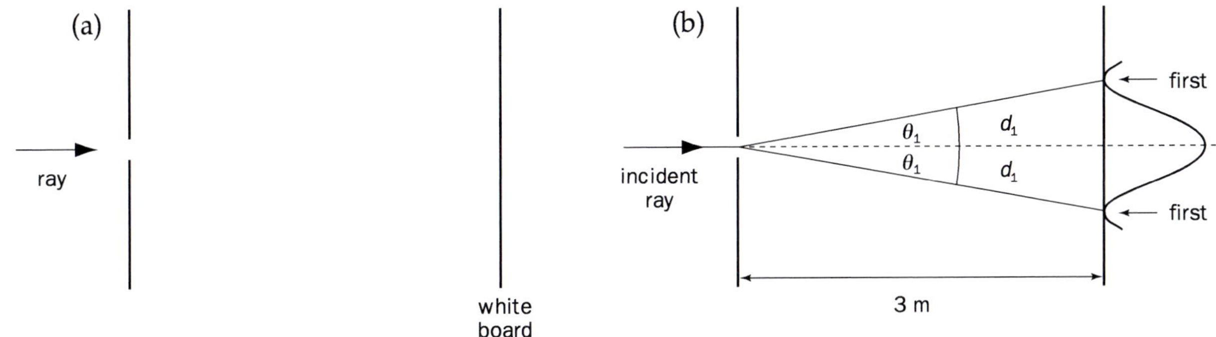

Double-slit experiment

BACKGROUND INFORMATION

In 1801, Thomas Young performed the following experiment, which provided evidence of the wave nature of light.

He had been trying to get two beams of light to interfere with each other. Firstly he used white light and placed a filter in front of the source to absorb all the wavelengths except one. He then sent the monochromatic beam through two slits, thus splitting the wavefront. Although the beams were monochromatic and made up of light of the same amplitude, he did not observe interference patterns. This was because the beams were not coherent. Eventually, he hit on the idea of having a single slit after the filter, which diffracted the light as a coherent beam on to the double slit. He observed on the screen a uniform pattern of bright and dark bands called fringes. All waves can experience superposition, whatever their type. The name interference is given to the pattern produced by the interaction. The bright fringes are the result of light superimposing constructively, and the dark bands indicate destructive interference (see Figure 1).

Figure 1

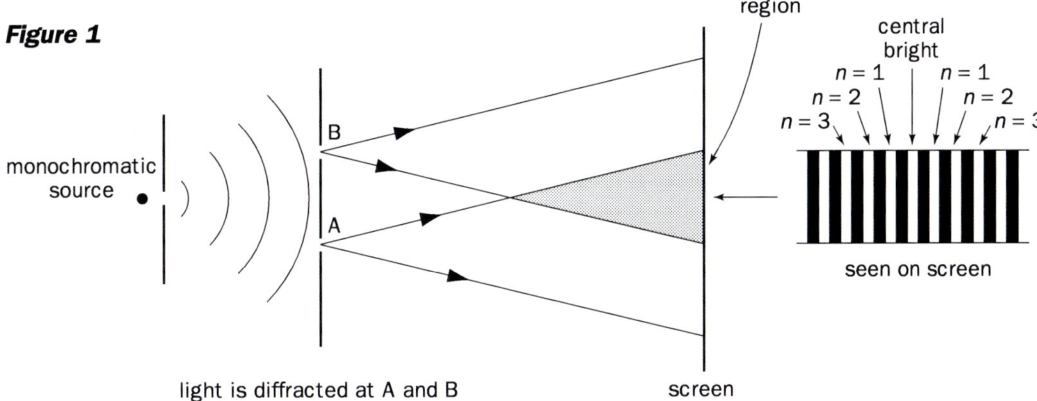

The central fringe is bright as both beams have travelled the same distance to reach that point. There are dark fringes on either side of the centre and these are the result of the path difference between the waves being half a wavelength. The mathematical relationship between wavelength, slit separation and fringes is given using the symbols shown in Figure 2.

Figure 2

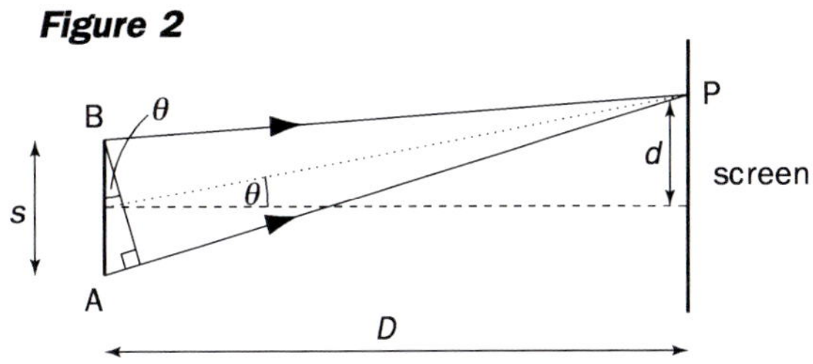

Let s be the distance from the centre of one slit to the centre of the second. D is the distance from slits to screen, and d the distance from the central bright fringe to the fringe being measured. For bright fringes to form, the path difference between the two rays, AP and BP, must be n, where $n = 0, 1, 2, 3\ldots$

$$\sin\theta = \frac{\mathrm{AP} - \mathrm{BP}}{s} \quad \text{and} \quad \tan\theta = \frac{d}{D}$$

But $\tan\theta \approx \sin\theta$ for small angles, and we can say $\frac{d}{D} = \frac{\mathrm{AP} - \mathrm{BP}}{s}$.

For bright fringes $\frac{ds}{D} = n\lambda$. For dark fringes $\frac{ds}{D} = (n + \frac{1}{2})\lambda$.

6 Double-slit experiment

USES OF INTERFERENCE

Compact discs (CD) have many tracks very closely spaced and these act as a diffraction grating. The white light falling on to the surface is diffracted and different colours are observed. The surface of the CD is coated with reflecting aluminium. The CD is read by interference. Each track has a series of pits in its surface, each a quarter of a wavelength thick. Light from the laser (which is coherent) is shone up on to the track and reflected back to the amplifier. When the light is reflected at the edge of a pit the reflected beam interferes destructively with the beam beside it, and no reflected beam reaches the amplifier. The signal changes from 1 to 0 (see Figure 3).

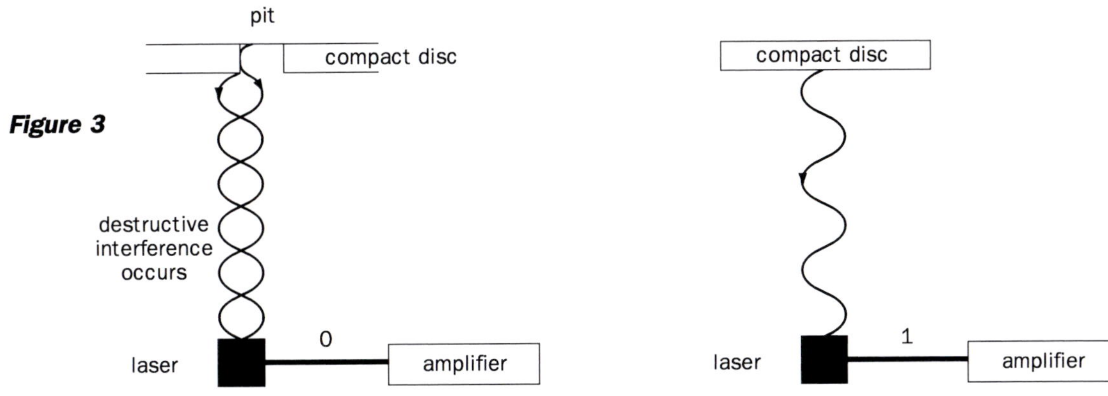

Figure 3

Camera lenses are coated with magnesium fluoride to a thickness of a quarter of a wavelength of one particular frequency. This reduces the amount of light that is reflected from the camera lens.

Checking glass surfaces for smoothness involves shining a sodium lamp on to the piece of glass to be checked. The glass block is mounted on a second block, which is the standard, and there is a layer of air between the two blocks. Interference rings (Newton's rings) are observed and, if both surfaces are smooth, the rings are evenly spaced. If the surfaces are uneven, the interference fringes appear like the contours on a map. In this case, division of amplitude produces the fringes.

The ray is partially reflected from the first boundary and partially from the second boundary. The light at the second boundary has undergone a phase change because of the boundary density (see Figure 4).

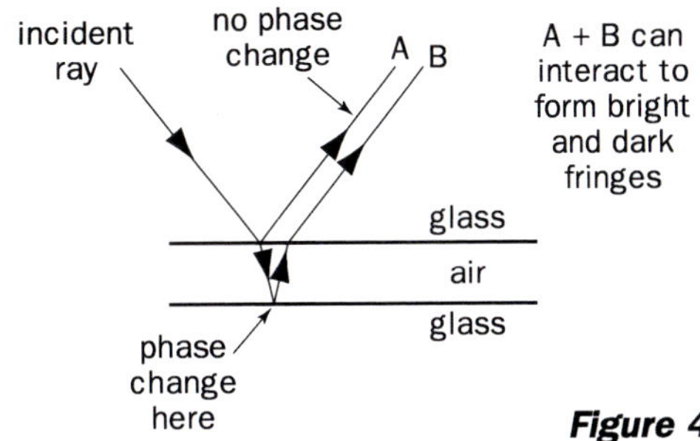

Figure 4

Double-slit experiment

LESSON POINTERS

The ripple tank demonstration with two slits is an excellent way to introduce this topic. All students should have the opportunity to see a double-slit experiment with both monochromatic and white light. The appearance of the fringes gives the teacher an opportunity to point out colour mixing. Magenta will be observed where red and blue fringes overlap. (Green light will have experienced destructive interference at this point.)

The mathematics for showing where the equation comes from can be done in a number of ways; I have used one that makes the assumption of $\sin\theta \approx \tan\theta$ for small angles. I have not discussed missing orders or the effect of a diffraction pattern and interference pattern being superimposed on each other. These are topics that lend themselves to class discussion, and each teacher knows their own group and how far to take matters.

Demonstrations of double-slit experiments should include microwaves and ultrasound to show that this holds for all waves.

It is interesting to work through a Lloyd's mirror problem with a class, to show them what happens when you have a virtual image as one slit. The phase change on reflection produces dark fringes where you would expect bright ones and vice versa.

ANSWERS TO WORKSHEET

1 (a) The oil forms a layer only a few molecules thick on the surface of the water. Some of the incident light is reflected off the top surface and undergoes a phase change of π. The rays, which are reflected off the bottom of the oil surface, undergo no phase change on reflection. The two emerging rays interfere with each other and, depending on the thickness of the oil film, certain wavelengths are missing from the reflected beam.
(b) The reflected light from the CD has certain wavelengths missing because of interference.
(c) Pearl shells are made up of thin layers of translucent material. Light, which is reflected from the surfaces, produces interference patterns in the same way oil films do.

2 Wave energy from region of destructive interference is redistributed to regions of constructive interference.

3 The diagram should show reflection from the upper and lower surfaces. A phase change occurs at the top surface only, so destructive interference takes place when the path difference is a whole number of wavelengths.

4 0.762 m

5 (a) 0.675 mm
(b) Students should mention the inverse square law.

6 (a) 0.900 mm
(b) 0.356 mm
(c) 0.178 mm
(d)

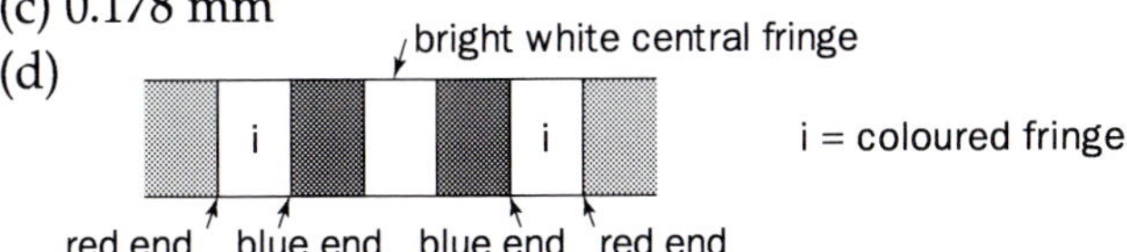

Double-slit experiment

BASIC FACTS

- In 1801, Thomas Young performed the double-slit experiment, which provided evidence of the wave nature of light. A uniform pattern of bright and dark bands appeared on the screen.

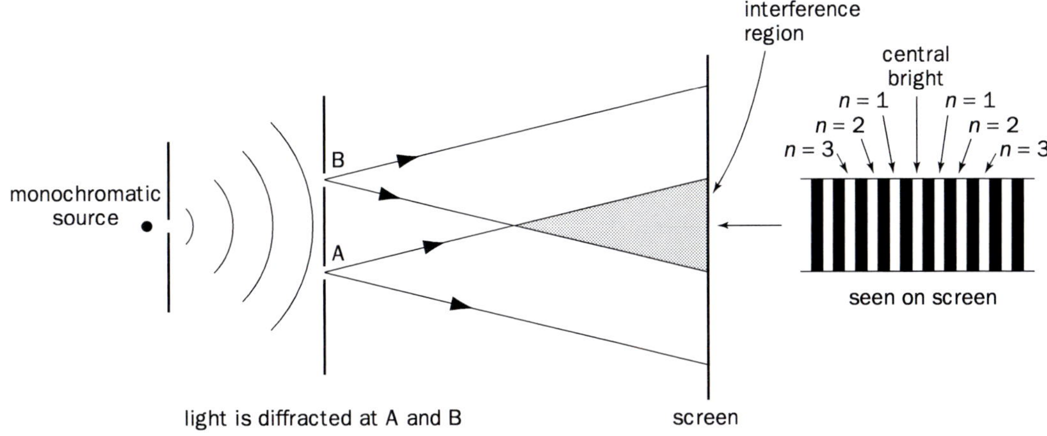

- All waves can experience superposition, whatever their type. The name interference is given to the pattern produced by the interaction. The bright fringes are the result of light superimposing constructively, and the dark bands indicate destructive interference
- For constructive interference, the path difference between two waves must be a whole number of wavelengths.
- For bright fringes $\frac{ds}{D} = n\lambda$. For dark fringes $\frac{ds}{D} = (n + \frac{1}{2})\lambda$.

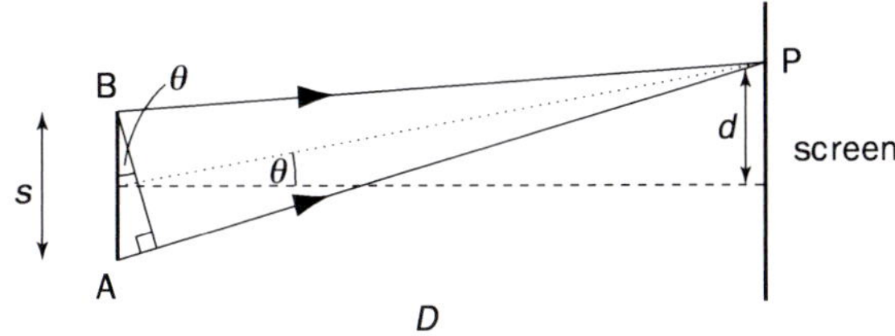

- The same fringe pattern is observed with white light, but white light is made up of many wavelengths. Each wavelength produces its own interference pattern, and these can overlap. The central fringe is made up of all the wavelengths and so is white.

Uses of interference

- Compact discs (CD) have many tracks very closely spaced and these act as a diffraction grating. The white light falling on to the surface is diffracted and different colours are observed. The surface of the CD is coated with reflecting aluminium. The CD is read by interference. Each track has a series of pits in its surface, each a quarter of a wavelength thick. Light from the laser (which is coherent) is shone up on to the track and reflected back to the amplifier. When the light is

Double-slit experiment

reflected at the edge of a pit the reflected beam interferes destructively with the beam beside it, and no reflected beam reaches the amplifier. The signal changes from 1 to 0 (see diagrams).

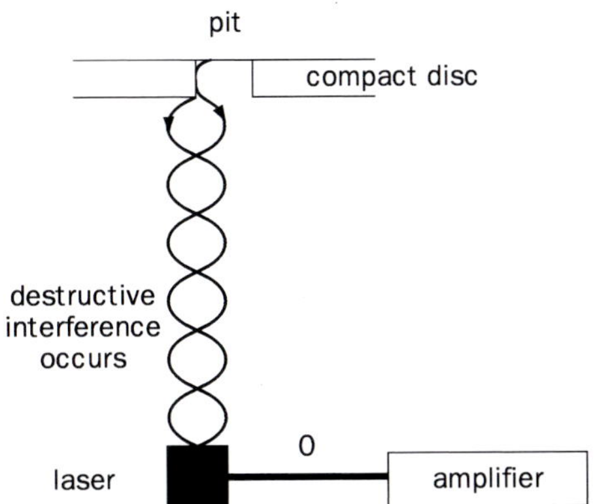

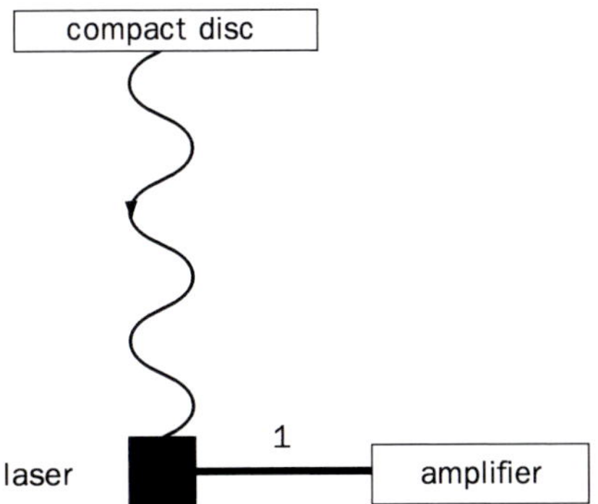

- Camera lenses are coated with magnesium fluoride to a thickness of a quarter of a wavelength of one particular frequency. This reduces the amount of light that is reflected from the camera lens.
- Checking glass surfaces for smoothness involves shining a sodium lamp on to the piece of glass to be checked. The glass block is mounted on a second block, which is the standard, and there is a layer of air between the two blocks. Interference rings (Newton's rings) are observed and, if both surfaces are smooth, the rings are evenly spaced. If the surfaces are uneven, the interference fringes appear like the contours on a map. In this case, division of amplitude produces the fringes. The ray is partially reflected from the first boundary and partially from the second boundary. The light at the second boundary has undergone a phase change at the boundary.

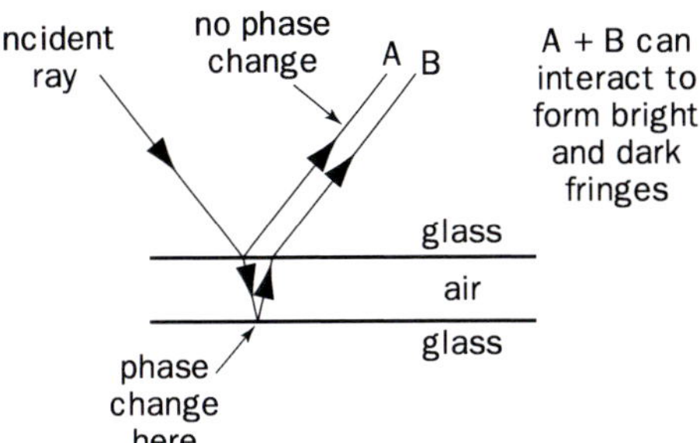

Double-slit experiment

QUESTIONS

1 (a) What causes the colour of an oil film on water?
(b) What causes the colour of a CD?
(c) What causes the colour of a pearl shell?

2 When two waves interfere destructively with each other, where does the wave energy go?

3 Complete the diagram on the right to show how interference can occur. What must be the path difference (in wavelengths) for there to be destructive interference?

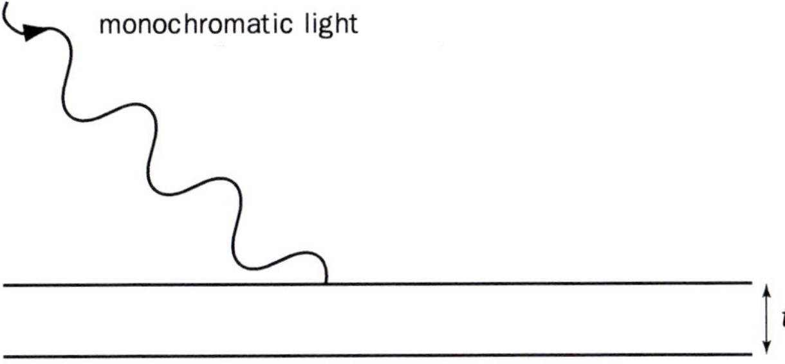

4 You set up a neon laser (λ = 700 nm) in a Young's double-slit experiment. You measure the distance from the first to the tenth bright fringe and find it to be 8.00 mm. If the slit separation is 0.60 mm, find the distance from the slits to the screen.

5 (a) You use a light of wavelength 540 nm to produce fringes on a screen 0.5 m away from the slits. If the slit separation is 0.40 mm, calculate the fringe separation.
(b) Why does the intensity of the fringe decrease on either side of the central bright fringe?

6 (a) Yellow light of wavelength 600 nm is used to illuminate a double-slit and fringes are observed 0.8 m from the slits. The distance from the first to the tenth fringe is 4.8 mm. Calculate the slit separation.
(b) The yellow light is replaced with blue light of wavelength 400 nm. What is the fringe separation now?
(c) You move the screen to 0.4 m from the slits. What is the spacing of the blue fringes now?
(d) You now replace the blue light with white light (range 400 nm–700 nm). Sketch what you would observe.

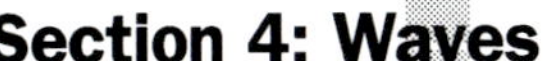

7 Stationary waves

BACKGROUND INFORMATION

Stationary waves are a special case of interference and are produced when two waves with the same amplitude, frequency and speed, but travelling in opposite directions, are superimposed. Longitudinal and transverse waves can produce standing waves. Unlike progressive waves, which transmit energy, stationary waves contain the energy within the system.

Look at Figure 1, which shows a set-up for producing stationary or standing waves. The *nodes* have no amplitude; other parts of the wave have maximum amplitude and these are called *antinodes*. Stationary waves are usually produced by the reflection of a progressive wave at a boundary.

Figure 1

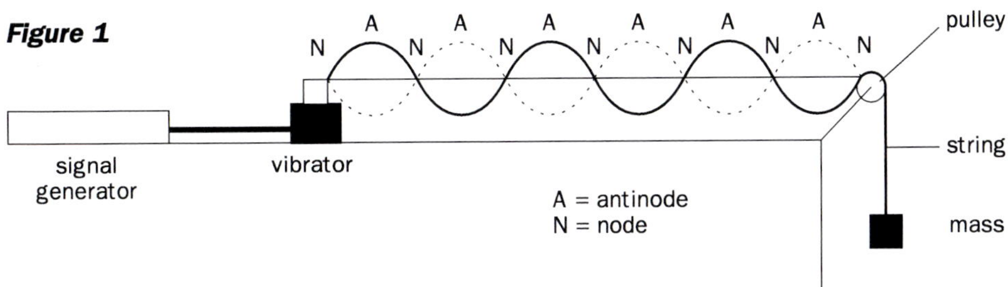

All the particles between two nodes are in phase with each other. The distance between two nodes is $\lambda/2$ and the maximum resultant amplitude is the sum of the amplitudes of the two waves. The nodes remain stationary.

You will have observed that, when the apparatus shown above is used, the string shows certain patterns at particular frequencies. These patterns are shown in Figure 2, and you can see that the distance between adjacent nodes is half a wavelength.

Figure 2

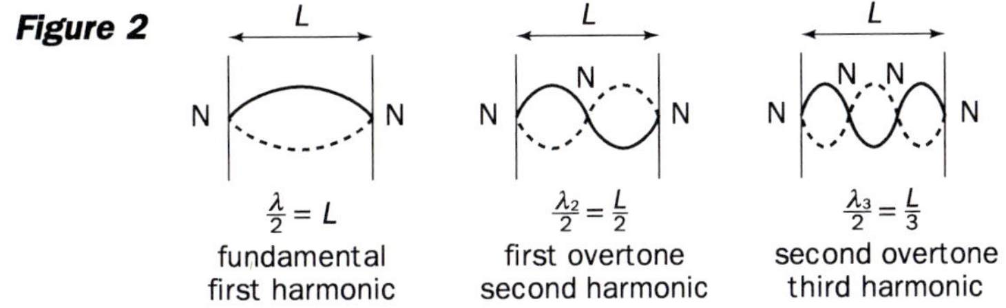

The formation of a harmonic depends on the tension T in the string (in N) and the mass per unit length of the string (in kg m^{-1}), which we will call μ. The equation $v^2 = T/\mu$ shows the relationship these have with the velocity of the wave in the string. The velocity v is related to the frequency of the nth harmonic f_n, by the equation $v = \lambda_n \times f_n$, and from the above we can see that $\lambda_n/2 = L/n$. Therefore,

$$f_n = \frac{n}{2L}\sqrt{\frac{T}{\mu}}$$

7 Stationary waves

STATIONARY WAVES IN PIPES

(a) Pipes closed at one end

The air at the closed end cannot move and this means that a node is formed at that point.

An antinode is formed at the open end. The fundamental (first harmonic) is shown in Figure 3, together with the first and second overtones. Why can we not produce the second and fourth harmonic?

Figure 3

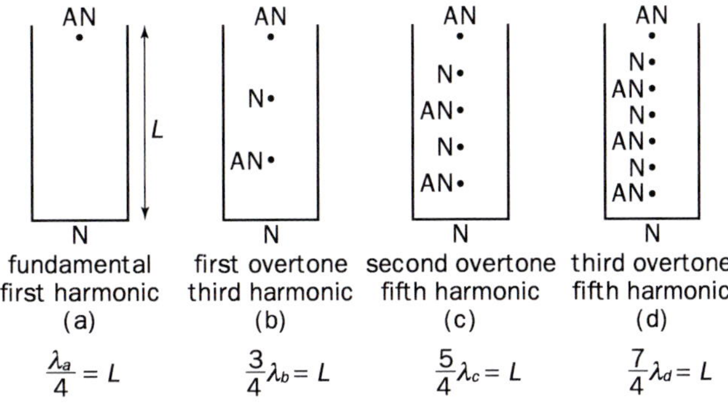

(b) Pipes open at both ends

Antinodes are produced at both ends of open pipes and, as you can see from Figure 4, all harmonics are present.

Figure 4

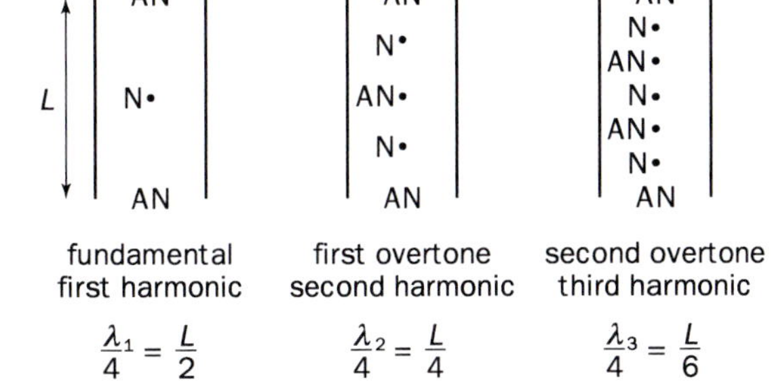

LESSON POINTERS

We have looked at superposition in terms of diffraction and interference. However, stationary waves and beats are two further examples of this phenomenon.

This section will look at standing waves, and pupils who are interested can look up the conditions for producing beats in their textbooks.

The classic demonstration is shown on the student sheet, and this serves as very good introduction to the topic. When using the stroboscope to view the waves, care should be taken to identify pupils with epilepsy or those who suffer from headaches. The stroboscope should be on for the shortest time possible.

Many musical instruments form stationary waves, and pupils who play an instrument could be encouraged to bring it into the laboratory. This gives all students the opportunity to appreciate where the boundaries are and the length of material vibrating.

Stationary waves

Making notes by blowing over test tubes filled with different amounts of water demonstrates resonance and standing waves. Ridged plastic tubing can be purchased which, when rotated horizontally or vertically, produces various sounds. These visual experiments help pupils to remember this topic.

A standing microwave wave experiment can be done later when the students could work out themselves where they expected nodes and antinodes to be. A loudspeaker connected to a signal generator can be used to investigate fundamentals and overtones in open and closed-at-one end tubes. A large measuring cylinder can be partially filled with water and an open-ended glass tube can be placed inside it. A tuning fork is used to obtain resonances in the air column above the water, and moving the inner glass tube up and down can alter the length of the column of vibrating air.

A question is raised in the paragraph on pipes closed at one end about the second and subsequent even harmonics not being formed. Some pupils will need to sketch several harmonics to convince themselves of this fact.

ANSWERS TO WORKSHEET

1

Stationary waves	**Progressive waves**
All the particles between two nodes have the same phase.	All the particles in one wavelength have a different phase.
Energy remains in the closed system.	Energy is transferred with the wave.
The amplitude depends on position.	The amplitude is the same for all particles through which the wave travels.
All the particles vibrate with the same frequency as the wave except the nodes which are at rest.	All the particles vibrate with the same frequency as the wave.

2 219.3 Hz; 877.2 Hz

3 3.0 cm; 12.0 cm

4 0.165 m

5 0.32 m; 768 Hz; 512 Hz; 1024 Hz

6 91.8 cm; 313 m s^{-1}; 171 Hz

7 Stationary waves

BASIC FACTS

- Stationary waves are a special case of interference and are produced when two waves with the same amplitude, frequency and speed, but travelling in opposite directions, are superimposed.
- Longitudinal and transverse waves can produce standing waves. Unlike progressive waves, which transmit energy, stationary waves contain the energy within the system.
- The diagram shows a set-up for producing stationary or standing waves. The *nodes* have no amplitude; other parts of the wave have maximum amplitude and these are called *antinodes*. Stationary waves are usually produced by the reflection of a progressive wave at a boundary.

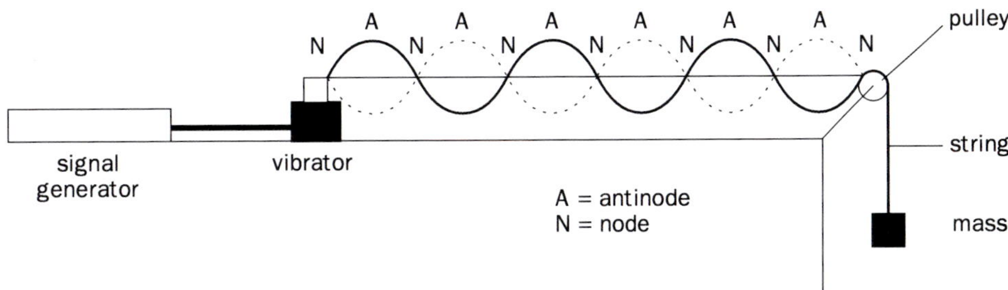

- All the particles between two nodes are in phase with each other. The distance between two nodes is $\lambda/2$ and the maximum resultant amplitude is the sum of the amplitudes of the two waves. The nodes remain stationary.
- You will have observed that, when the apparatus shown above is used, the string shows certain patterns at particular frequencies. These patterns are shown in the diagram below, and you can see that the distance between adjacent nodes is half a wavelength.

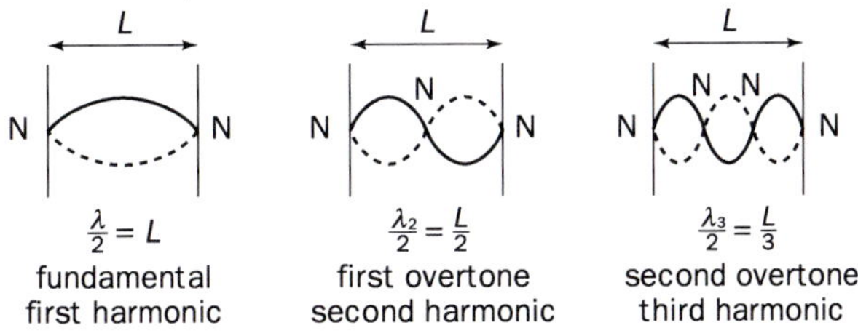

- The formation of a harmonic depends on the tension T in the string (in newtons) and the mass per unit length μ of the string (in kg m^{-1}).
- $v^2 = T/\mu$ shows the relationship these have with the velocity of the wave in the string.
- The velocity v is related to the frequency of the nth harmonic f_n, by the equation $v = \lambda_n \times f_n$, and from the above we can see that $\lambda_n/2 = L/n$ and $f_n = \frac{n}{2L}\sqrt{\frac{T}{\mu}}$.

Stationary waves in pipes

(a) Pipes closed at one end

- The air at the closed end cannot move and this means that a node is formed at that point.

Stationary waves

- An antinode is formed at the open end. The fundamental (first harmonic) is shown in the diagram, together with the first and second overtones.

(b) Pipes open at both ends

- Antinodes are produced at both ends of open pipes and, as you can see from the diagrams, all harmonics are present.

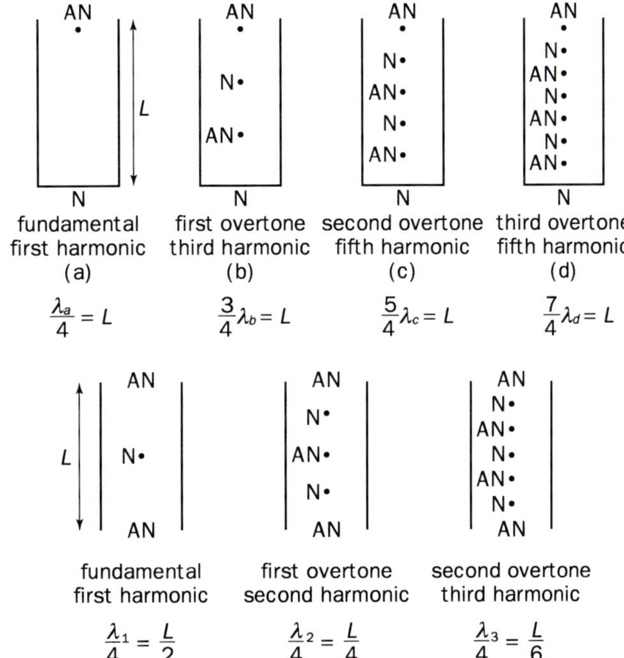

QUESTIONS

1 Compare and contrast a stationary wave and a progressive wave.

2 A string of length 80 cm has a mass of 1.3 g and is under tension of 200 N. What is the fundamental frequency? What is the frequency of the fourth harmonic?

3 A microwave transmitter emits waves of frequency 1×10^{10} Hz. It faces a metal plate, and a microwave detector moves between transmitter and plate. Calculate the wavelength of the microwaves. How far would the detector have to move from the first to the ninth consecutive node from the metal screen? Assume the speed of microwaves in air is 3×10^8 m s^{-1}.

4 Two loudspeakers are set up facing each other at a local fete, and a loud feedback signal of frequency 1000 Hz is transmitted through both speakers. If the speakers are 24 m apart, calculate the distance apart of successive nodes. Assume the speed of sound in air is 330 m s^{-1}.

5 A pipe, which is closed at one end, has a fundamental frequency of 256 Hz. What is the length of the pipe? What is the frequency of the first overtone? What would be the fundamental frequency and that of the first overtone be if the pipe was open at both ends? Assume the speed of sound in air is 330 m s^{-1}.

6 A vibrator is attached to one end of a horizontal wire, and the other end of the wire goes over a pulley to a mass of 10 kg. A note of frequency 512 Hz is produced when the following pattern is observed.

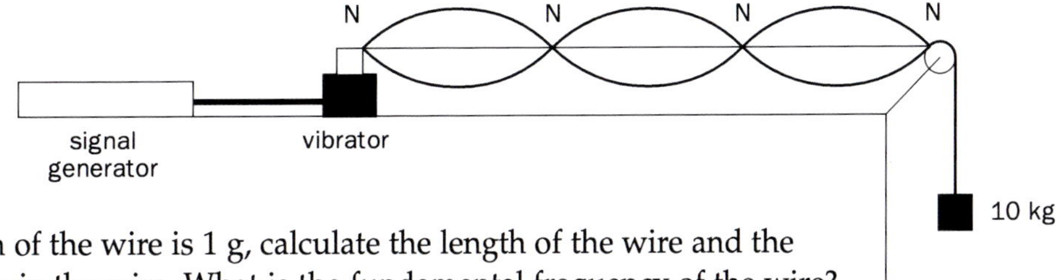

If the mass of 1 m of the wire is 1 g, calculate the length of the wire and the speed of the wave in the wire. What is the fundamental frequency of the wire?

SECTION
5
Nuclear and Atomic Physics

Equivalence of mass and energy

BACKGROUND INFORMATION

One of the most important consequences of Einstein's special theory of relativity is that mass and energy are equivalent. The theory predicts that, if a body gains an amount of energy E, it will increase its mass by an amount m, given by

$$E = mc^2$$

where c is the velocity of light.

The converse of this statement is also true, i.e. a decrease in the mass results in the release of energy.

Two fairly simple examples of the power of this equation could be shown.

EXAMPLE 1

What is the increase in mass of 1 kg of water when its temperature is raised from 20°C to 100°C?

$$E = 1\ \text{kg} \times 4200 \frac{\text{J}}{\text{kg K}} \times 80\ \text{K} = 336\ 000\ \text{J}$$

$$m = 336\ 000/(9 \times 10^{16}) = 3.7 \times 10^{-12}\ \text{kg}$$

i.e. hardly measurable.

EXAMPLE 2

How much energy would be released if it were possible to disintegrate completely 1 g of chalk?

$$E = 1 \times 10^{-3} \times 9 \times 10^{16} = 9 \times 10^{13}\ \text{J}$$

The size of this answer does not always mean a great deal to a student but, if you consider that a 1 million tonne supertanker has a weight of 10^{10} N, using *work* = *force* × *distance*, 9×10^{13} J is sufficient to lift the tanker 9 km into the air!

We are more interested in using the equation on the nuclear level than on the macro scale of the examples above.

In radioactive decay, nuclear fission and nuclear fusion, there is always a net decrease in mass and therefore a release of energy. But before considering nuclear reactions, it is necessary to look at two other useful and important concepts.

Equivalence of mass and energy

The **unified atomic mass unit**, u, is defined as one twelfth of the mass of a ^{12}C atom. Thus, since 12 g of ^{12}C contains 6.02×10^{23} atoms, 1 atom of carbon has a mass of $\frac{12}{6.02 \times 10^{26}}$ kg and

$$1\ u = \frac{12}{12 \times 6.02 \times 10^{26}} kg = 1.66 \times 10^{-27}\ kg$$

An **electron volt** is the energy gained by an electron falling through a potential difference (p.d.) of 1 V.

Since *energy* = *p.d.* × *charge*
1 eV = 1.6×10^{-19} J and
1 MeV (a million electron volts) = 1.6×10^{-13} J.
(This is a useful quantity in nuclear physics.)

Example 2 above shows that a decrease in mass of 1 kg would produce 9×10^{16} J and a simple calculation shows that 1 u = 931 MeV. Now use an example from radioactivity to bring all of this together.

EXAMPLE 3

When radon-222 emits an alpha particle, there is a decrease in mass of 0.0052 u. Calculate the energy released (a) in J and (b) in MeV.

Mass decrease $= 0.0052 \times 1.66 \times 10^{-27}$ kg
Energy released $= 0.0052 \times 1.66 \times 10^{-27} \times (3 \times 10^{8})^2 = 7.77 \times 10^{-13}$ J
$= 7.77 \times 10^{-13}/1.6 \times 10^{-13} = 4.85$ MeV

ANSWERS TO WORKSHEET

1 1.3×10^{-30} kg

2 (a) 1.5×10^{-10} J
(b) 930 MeV

3 1.34×10^{-12} J

4 1.4×10^{3} J m^{-2}

Equivalence of mass and energy

student WORKSHEET

BASIC FACTS

- One of the most important consequences of Einstein's special theory of relativity is that mass and energy are equivalent. The theory predicts that, if a body gains an amount of energy E, it will increase its mass by an amount m given by $E = mc^2$, where c is the velocity of light.
- The converse of this statement is also true, i.e. a decrease in the mass results in the release of energy.
- The unified atomic mass unit, u, is defined as one-twelfth of the mass of a ^{12}C atom. Thus, since 12 g of ^{12}C contains 6.02×10^{23} atoms, 1 atom of carbon has a mass of $\frac{12}{6.02 \times 10^{26}}$ kg and

$$1 \text{ u} = \frac{12}{12 \times 6.02 \times 10^{26}} \text{ kg} = 1.66 \times 10^{-27} \text{ kg}$$

- An electron volt is the energy gained by an electron falling through a potential difference (p.d.) of 1 V.
- Since *energy = p.d. × charge*,

  1 eV = 1.6×10^{-19} J and

  1 MeV (a million electron volts) = 1.6×10^{-13} J. (This is a useful quantity in nuclear physics.)

QUESTIONS

1 Lead-209 is a radioisotope of lead, which emits beta particles with an energy of 0.72 MeV. Assuming all of the energy released is given to the beta particle, calculate the loss in mass when decay occurs.

2 Calculate the energy released when 1 u is converted into energy. Calculate your answer in (a) J and (b) MeV.

3 When polonium-206 emits an alpha particle, there is a loss in mass of 0.009 u. Calculate the energy released.

4 Imagine the Earth to be part of a sphere of radius 1.35×10^{11} m, with the Sun at the centre. The Sun converts about 4×10^6 tonnes of matter per second into energy. Calculate the intensity of energy reaching the Earth from the Sun. (Intensity is power per square metre.)

Mass defect and binding energy

BACKGROUND INFORMATION

The mass of any atom or nucleus is always less than the sum total of the masses of its constituents.

For helium, a helium atom is made of 2 protons, 2 neutrons and 2 electrons. The mass of a proton is 1.007 28 u, that of a neutron is 1.008 66 u, and of an electron 0.000 55 u.

The total mass of constituents = 2(1.007 28 + 1.008 66 + 0.000 55) u = 4.032 98 u.
The measured mass of a helium atom is 4.002 60 u. Thus, there is a difference in mass of 0.030 38 u. This difference in mass is known as the *mass defect*.

This mass disappears when the nucleus is formed. Einstein showed that there is a mass–energy equivalence, and the mass defect is equivalent to the amount of energy released on the formation of the nucleus from its constituent particles. This energy released is known as the *binding energy* of the nucleus. Alternatively, it is the energy required to separate completely the nucleons in a nucleus.

Again for helium: it was shown in the previous lesson that 1 u = 931 MeV. Thus, the binding energy for a helium nucleus is 0.030 38 × 931 = 28.3 MeV.

This calculation can also be carried out in joules:
1 u = 1.66×10^{-27} kg
0.030 38 u = $0.030\,38 \times 1.66 \times 10^{-27} = 5.04 \times 10^{-29}$kg
Using $E = mc^2$:
binding energy = $5.04 \times 10^{-29} \times (3 \times 10^8)2 = 4.53 \times 10^{-12}$ J
(*Note:* 1 eV = 1.6×10^{-19} J and, therefore, 28.3 MeV = $1.6 \times 10^{-19} \times 28.3 \times 10^6 = 45.3 \times 10^{-12}$ J.)

ANSWERS TO WORKSHEET

1 (a) 0.2207 u; 3.3×10^{-11} J or 205 MeV
(b) 0.030 38 u; 4.5×10^{-12} J or 28.3 MeV
(c) 0.369 08 u; 5.5×10^{-11} J or 344 MeV

2 59 700 kWh

Mass defect and binding energy

student WORKSHEET

BASIC FACTS

- The mass of any atom or nucleus is always less than the sum total of the masses of its constituents.
- This difference in mass is known as the mass defect.
- This mass disappears when the nucleus is formed. Einstein showed that there is a mass–energy equivalence, and the mass defect is equivalent to the amount of energy released on the formation of the nucleus from its constituent particles. This energy released is known as the binding energy of the nucleus. Alternatively, it is the energy required to separate completely the nucleons in a nucleus.
- The binding energy is often measured in MeV and to convert into joules

  1 MeV = 1.6×10^{-13} J

- The mass defect is often measured in unified atomic units u and to convert into kg

  1 u = 1.66×10^{-27} kg

- The binding energy can be calculated using $E = mc^2$, where E is the binding energy in joules and m is the mass defect in kilograms.
- Alternatively 1 u = 931 MeV can be used to convert directly from mass into energy.

QUESTIONS

In the following question use: mass of a proton = 1.007 28 u, mass of a neutron = 1.008 66 u, mass of an electron = 0.000 55 u.

1 Calculate the mass defects and binding energies, in joules and MeV, for the following:

(a) a magnesium atom consisting of 12 electrons, 12 protons and 13 neutrons. The mass of the atom is 24.985 84 u;

(b) a helium atom of mass 4.002 60 u;

(c) an argon-40 atom of mass 39.962 38 u and proton number 18.

2 Calculate the energy, in kilowatt-hours, which would be produced if a mole of protons, a mole of neutrons and a mole of electrons could be combined to produce a mole of heavy hydrogen atoms, each of mass 2.01410 u. ($N_A = 6.02 \times 10^{23}$)

Fusion

BACKGROUND INFORMATION

The graph of binding energy in Figure 1 shows a steady increase in the binding energy per nucleon as the mass number increases up to 56. Beyond that, it shows a slight decrease as the mass number increases further. One anomaly to the pattern is the position of helium, and this can be explained as due to the great stability of the helium nucleus.

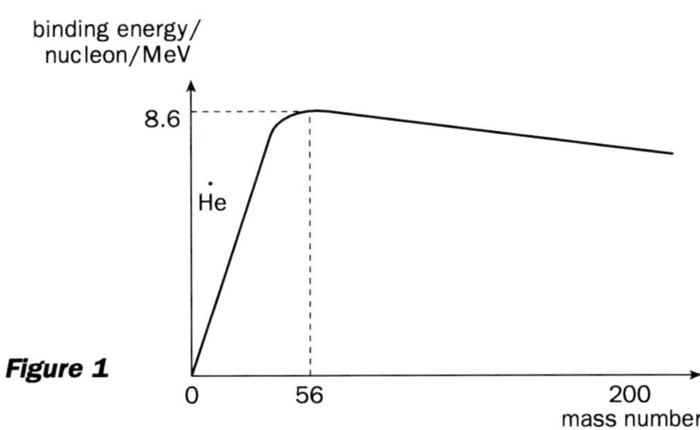

Figure 1

The greater the binding energy per nucleon, the more stable is the nucleus because of the greater energy per particle that has to be supplied to separate the nucleus into its constituents.

Fusion is the process whereby two small nuclei come together to form a larger nucleus.

It can be seen from Figure 1 that if two light nuclei, say $A < 20$, fused together, the binding energy per nucleon of the resulting nucleus would be greater than that of either of the original nuclei. Following through the logic of this argument, the new nucleus must be more stable than the original nuclei, and energy must therefore be given off in the fusion process. The energy given off is equivalent to the loss in mass when comparing the total final mass to the total mass of the original nuclei.

EXAMPLE

Many fusion reactions take place in the Sun. One possible reaction is the fusion of a deuterium nucleus with a tritium nucleus to form a helium nucleus and a neutron.

$${}^{3}_{1}\mathrm{H} + {}^{2}_{1}\mathrm{H} \rightarrow {}^{4}_{2}\mathrm{He} + {}^{1}_{0}\mathrm{n}$$

To calculate the energy released from such a reaction, we need to know the following data:
mass of deuterium nucleus = 2.014 19 u
mass of tritium nucleus = 3.016 46 u
mass of helium nucleus = 4.002 77 u
mass of neutron = 1.008 98 u

Total mass on left-hand side of equation = 3.016 46 u + 2.014 19 u = 5.030 65 u
Total mass on right-hand side of equation = 4.002 77 u + 1.008 98 u = 5.011 75 u
Loss in mass = 5.030 65 u – 5.011 75 u = 0.0189 u
Now 1 u = 931 MeV
Energy released = 0.0189 × 931 = 17.6 MeV
or
$17.6 \times 10^{6} \times 1.6 \times 10^{-19} = 28.2 \times 10^{-13}$ J

3 Fusion

LESSON POINTERS

The graph of binding energy per nucleon against mass number is a useful starting point when teaching about fusion and fission.

The example can be worked through by the teacher and students.

ANSWERS TO WORKSHEET

1 (i) A nucleus containing 6 protons and 7 neutrons
(ii) Energy released in (a) = 2.52×10^{-12} J, energy released in (b) = 2.83×10^{-12} J; therefore (b) as (a) is too small.

2 6.0×10^{-13} J; 9×10^{13} J

3 Fusion

student WORKSHEET

BASIC FACTS

- A graph of binding energy per nucleon against mass number for common isotopes produces the shape in the diagram:

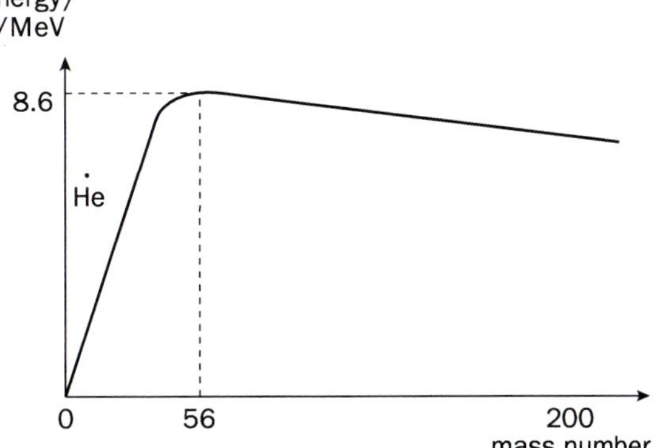

The shape of the graph is very important. It shows a steadily increasing binding energy per nucleon up to a mass number of 56 (iron), and then it slowly decreases as the mass number increases further.

- The significance of the graph is that it can be used to explain energy release in both fusion and fission reactions.
- Fusion is where two light nuclei come together to form a heavier nucleus.
- From the graph, if the original nuclei have mass numbers of less than say about 20, when they fuse the final nucleus will have a greater binding energy per nucleon than either of the original nuclei. This means that the final product is more stable than the original nuclei because it will take more energy to split it into its constituent particles than is required to split the original nuclei.
- Now if the final nucleus requires more energy to split it into its constituent particles, energy must have been released on the fusion of the original nuclei. This energy comes from a loss in mass when the nuclei fuse together.
- Fusion is only likely to occur in light nuclei, i.e. to the left of the maxima at $A = 56$, because of the increase in the binding energy per nucleon, resulting in the final nucleus becoming more stable.

QUESTIONS

1 When beryllium is bombarded with alpha particles of energy 8.0×10^{-13} J, carbon atoms are produced, together with a penetrating radiation of energy greater than 2.7×10^{-12} J. The nuclear reaction might be either (a) ${}^{9}_{4}\text{Be} + {}^{4}_{2}\text{He} \rightarrow {}^{13}_{6}\text{C}$ + gamma ray or (b) ${}^{9}_{4}\text{Be} + {}^{4}_{2}\text{He} \rightarrow {}^{12}_{6}\text{C} + {}^{1}_{0}\text{n}$.

(i) Explain what is meant by ${}^{13}_{6}\text{C}$.

(ii) Show that one of the reactions is possible and the other is not.

Nuclide	**Mass/u**
Neutron	1.0087
Be	9.0150
He	4.0040
C-13	13.0075
C-12	11.9967

2 If two deuterium nuclei fuse to form a helium-3 nucleus and a neutron, calculate the energy released. (Mass of neutron = 1.009 u, mass of deuterium = 2.015 u, mass of helium-3 = 3.017 u.) How much energy, in joules, would be released if 1 kg of deuterium were to fuse? ($1 \text{ u} = 1.66 \times 10^{-27}$ kg.)

Fission

BACKGROUND INFORMATION

In the Topic 3 on Fusion, the graph of binding energy per nucleon against mass number showed that, if light nuclei fuse together, there is a release of energy, resulting in a more stable nucleus and a greater binding energy per nucleon. Similarly, if a heavy nucleus could be induced to split into two smaller nuclei, they would occupy a position of higher binding energy per nucleon and be more stable.

Fission is the name given to the process in which a large nucleus splits into two smaller nuclei.

In 1938, Hahn and Strassmann were studying the effects of neutron bombardment of uranium when they discovered one of the products to be an isotope of barium. In the periodic table, barium (Z = 56) is so far removed from uranium (Z = 92) that it is not possible for it to be produced from a radioactive decay series of uranium. This reaction became the first reported fission reaction.

In a very short period of time the following facts were known:
(a) Uranium (Z = 92), thorium (Z = 90) and protactinium (Z = 91) could all be fissioned with neutrons.
(b) Large disintegration energies were released on fission.
(c) Fast neutrons were also emitted on fission.
(d) There was a range of fission products, all of which were radioactive and decayed by beta emission.

ENERGY RELEASE IN FISSION

In the fission referred to above, one reaction that could take place is represented by the equation

$${}^{235}_{92}\text{U} + {}^{1}_{0}\text{n} \rightarrow {}^{141}_{56}\text{Ba} + {}^{92}_{36}\text{Kr} + 3{}^{1}_{0}\text{n} + Q \text{ (energy released)}$$

Note that the superscripts on each side add up to 236, and the subscripts on each side add up to 92 (i.e. conservation of number of fundamental particles and of charge).

The masses of the constituents are as follows:

n = 1.008 665 u | Barium-141 = 140.914 050 u
Uranium-235 = 235.043 915 u | Krypton-92 = 91.920 564 u

Thus

total mass on left-hand side = 1.008 665 u + 235.043 915 u
= 236.052 58 u

total mass on right-hand side = 140.914 050 u + 91.920 564 u + 3.025 995 u
= 235.860 61 u

loss in mass = 236.05 258 u – 235.8606 u
= 0.191 975 u

energy released = 0.191 975 × 931 MeV
= 179 MeV.

The vast majority of this energy is in the form of kinetic energy (k.e.) of the fission fragments, and it is this energy that is used in a nuclear reactor to produce steam for driving the turbines. If one of the neutrons produced in the reaction goes on to fission with another uranium nucleus, a controlled chain reaction can be sustained.

4 Fission

TEACHER NOTES

ANSWERS TO WORKSHEET

1 (b) c = neutron; $x = 4$
(d) 210 MeV (as given)
(e) Reaction 1 more likely as final products must be in a lower energy state.

2 See graph Figure 1 Topic 3 Fusion (iii) 129.7 MeV

3 $1.56 \times 10^{17}\ \text{s}^{-1}$

4 1.68×10^{-10} J or 1044 MeV

Fission

BASIC FACTS

- Fission is the splitting up of a large nucleus into two smaller nuclei of approximately equal sizes.
- The graph from the last lesson suggests that if a large nucleus, i.e. one to the right-hand end of the graph, were to split in two, the smaller nuclei left would each have a greater binding energy per nucleon than the original atom. This means that they would be more stable, and therefore that energy must be released in the fission process.
- Fission was first shown to be possible in 1938, when Hahn and Strassmann bombarded uranium with neutrons and found that a radioactive isotope of barium was produced. Barium is in the middle of the periodic table, while uranium is towards the very end.
- Other features of fission reactions that are important to know:
  Uranium (Z = 92), thorium (Z = 90) and protactinium (Z = 91) can all be fissioned with neutrons.
  Large disintegration energies are released on fission.
  Fast neutrons are also emitted on fission.
  There are a range of fission products, all of which are radioactive and decay by beta emission.

Energy released in a fission reaction

- The energy released in a fission reaction is due to a loss in mass when the original nucleus fissions after absorbing the neutron.
- The fission reaction mentioned above can be represented by the following equation:

$$^{235}_{92}\text{U} + ^{1}_{0}\text{n} \rightarrow ^{141}_{56}\text{Ba} + ^{92}_{36}\text{Kr} + 3^{1}_{0}\text{n} + \text{Q (energy released)}$$

- Note that the superscripts on each side add up to 236, and the subscripts on each side add up to 92 (i.e. conservation of number of fundamental particles and of charge).
- The masses of the constituents are as follows:
  n = 1.008 665 u   Barium-141 = 140.914 050 u
  Uranium-235 = 235.043 915 u   Krypton-92 = 91.920 564 u
  We can calculate the total mass on the left- and right-hand sides of the equations and verify that the left-hand side is the greater.
  We can convert the mass difference into energy by using the relationship 1 u = 931 MeV.
- Sometimes the masses are given in kilograms. After finding the mass difference, the energy released can be found by using Einstein's relationship $E = mc^2$. The energy will then be given in joules.
- The vast majority of the energy given off is in the form of kinetic energy of the fission fragments, and it is this energy that is used in a nuclear reactor to produce steam to drive the turbines. Also, if one neutron from each reaction goes on to produce one further fission, it is possible to set up a controlled chain reaction.

Fission

QUESTIONS

Nuclide	A	Z	E/MeV
Kr	92	36	8.72
Sr	95	38	8.74
Ba	131	56	8.50
Xe	139	54	8.39
U	235	92	7.60

E = binding energy per nucleon

1 (a) Explain what is meant by
(i) the radioactive decay of a nucleus; (ii) nuclear fission.

When uranium-235 nuclei are fissioned by slow-moving neutrons, two possible reactions are:

reaction 1: ${}^{235}_{92}\text{U} + {}^{1}_{0}\text{n} \rightarrow {}^{139}_{54}\text{Xe} + {}^{95}_{38}\text{Sr} + 2{}^{1}_{0}\text{n}$ + energy

reaction 2: ${}^{235}_{92}\text{U} + {}^{1}_{0}\text{n} \rightarrow 2{}^{116}_{46}\text{Pd} + xc$ + energy

(b) For reaction 2, identify the particle c and state the number x of such particles produced in the reaction.

The binding energy per nucleon E for a number of nuclides is given above.
(c) What is meant by the binding energy per nucleon of a nucleus?
(d) Show that the energy released in reaction 1 is 210 MeV.
(e) The energy released in reaction 2 is 163 MeV. Suggest, with a reason, which one of the two reactions is more likely to occur.

2 Draw a graph to show the variation in binding energy per nucleon against the mass number of the nucleon.

Use the curve to explain why there is a release of energy when
(i) two light nuclei fuse to produce a heavier one (fusion)
(ii) a heavy nucleus disintegrates into two lighter ones (fission).

Uranium-235 can fission with slow neutrons to produce barium-131 and krypton-92.
(iii) Estimate the energy available using the information above.

3 The average fission reaction produces 200 MeV of energy. How many fissions per second are required for a 5 MW power station.

4 Uranium-235 can undergo neutron-induced fission to give the fission products strontium-90 and xenon-143. Using the following information, calculate the energy released.
Mass of neutron = 1.008 66 u
Mass of strontium-90 nucleus = 89.907 75 u
Mass of xenon-143 nucleus = 142.917 93 u
Mass of uranium-235 nucleus = 234.956 09 u

Wave–particle duality

TEACHER NOTES

BACKGROUND INFORMATION

In Topics 5–10 of Section 5 in the AS resource pack the idea of a photon of electromagnetic radiation is introduced. A photon has a quantum of energy, E given by

$$E = hf = hc/\lambda$$

where h is the Planck constant and f is the frequency. The quantum theory was developed from this idea and was originally necessary to explain spectra and the photoelectric effect. Since the early years of the last century the quantum theory has been developed and applied satisfactorily to explain and predict many phenomena. For instance, much of the behaviour of electrons in microchips can only be explained by the quantum theory and the micro-miniturisation of computer and mobile phone circuitry depends totally on quantum phenomena.

LESSON POINTERS

Since a photon has energy hf it also has momentum p. This is given by equating the energy of the photon to its mass using Einstein's $E = mc^2$ equation.

$$E = mc^2 = hc/\lambda$$

hence

$$p = mc = h/\lambda$$

This equation gives, for example, the force with which the Sun's light pushes on the Earth and it has been proposed that a spaceship might one day be pushed by sunlight, giving it a very small acceleration for a very long time.

In 1923 DeBroglie (pronounced DeBroy) suggested that since light has a dual nature, being considered a wave or a particle, perhaps particles could be considered as waves using the same theory, namely

$$p = h/\lambda$$

If values are substituted into this equation for large objects then the wavelength is very small and the wave-like nature of the object is impossible to detect, but if an electron is considered then the wavelength is larger and diffraction effects of electrons can be obtained. DeBroglie performed the following experiment in which he placed a thin film of graphite in the path of a beam of electrons and observed a circular diffraction pattern (Figure 1).

Figure 1

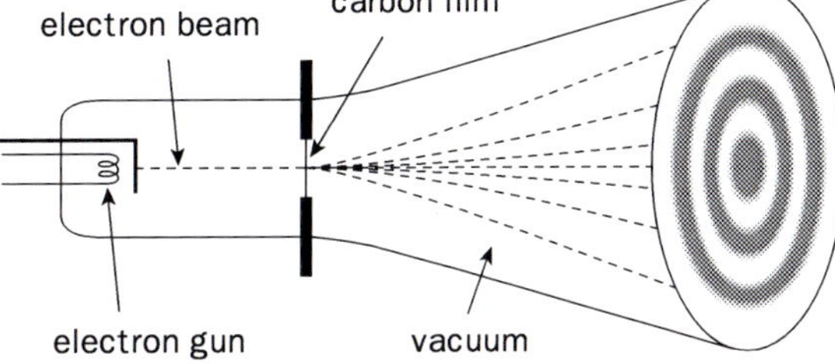

5 Wave–particle duality

TEACHER NOTES

The experiment can be repeated in the laboratory, using a piece of apparatus manufactured by Teltron, and clear diffraction rings can be observed. Since the separation of the atoms in the graphite is known, the wavelength of the electrons can be determined and good agreement with the equation is obtained.

ANSWERS TO WORKSHEET

1 1.32×10^{-27} N s

2 1.7×10^{19}

3 (a) 6.6×10^{-28} N s
(b) 7.1×10^{21}
(c) 4.7×10^{-6} N
(d) 4.7×10^{-6} N
(e) 4 800 000 m^2

4 2.3×10^{-11} m

5 8.0°, 16.2°

5 Wave–particle duality

STUDENT WORKSHEET

BASIC FACTS

- Last year, in your AS course, the idea of a photon of electromagnetic radiation was introduced. A photon has a quantum of energy, E, given by

  $E = hf = hc/\lambda$

  where h is the Planck constant and f is the frequency.

- Since a photon has energy hf it also has momentum p. This is given by equating the energy of the photon to its mass using Einstein's $E = mc^2$ equation.

  $E = mc^2 = hc/\lambda$

  hence

  $p = mc = h/\lambda$

- In 1923 DeBroglie (pronounced DeBroy) suggested that since light has a dual nature, being considered a wave or a particle, perhaps particles could be considered as waves using the same theory, namely

  $p = h/\lambda.$

  If values are substituted into this equation for large objects then the wavelength is very small and the wave-like nature of the object is impossible to detect, but if an electron is considered then the wavelength is larger and diffraction effects of electrons can be obtained.

- The following experiment involves putting a thin film of graphite in the path of a beam of electrons. A circular diffraction pattern is obtained because the atoms in the graphite act as a diffraction grating. As they are not aligned in just one direction the first order maximum will occur at a fixed angle to the electron beam, giving the circular pattern (Figure 1). The second order diffraction pattern can also be seen.

Figure 1

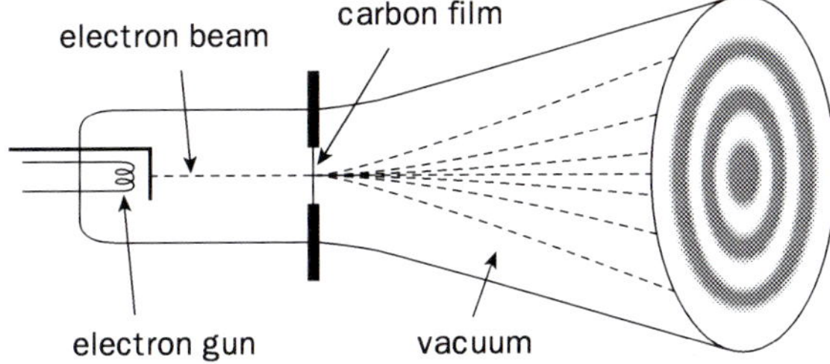

Wave–particle duality

student WORKSHEET

QUESTIONS

1 Calculate the momentum of a photon of light of wavelength 5.0×10^{-7} m.

2 How many photons of light are emitted every second from a 60 W light bulb of efficiency 11%? (This means that only 11% of 60 W is emitted in the form of light.) Assume that the average wavelength of the light emitted is 5.0×10^{-7} m.

3 On the Earth the Sun provides 1400 W of power per square metre. The range of wavelengths stretches from infrared to ultraviolet. Making the rather large assumption that the average wavelength is 1.0×10^{-6} m, calculate

(a) the momentum of one photon of this wavelength

(b) the number of photons striking one square metre of the Earth every second

(c) the rate of change of momentum of these photons, assuming they are all absorbed

(d) the force on each square metre of the vanes of a satellite at this distance from the Sun and facing the Sun

(e) the area of vanes required for the satellite, of mass 2800 kg, if it is to have an acceleration away from the Sun of 0.008 m s^{-2}.

4 Calculate the wavelength of an electron travelling at 3.2×10^7 m s^{-1}. The mass of an electron is 9.1×10^{-31} kg; the Planck constant is 6.6×10^{-34} J s.

5 An electron is accelerated by a potential difference of 3000 V before striking a layer of graphite in which the atoms are separated by a distance of 1.6×10^{-10} m. Calculate the angles of diffraction for both the first and second order diffraction circles. (Hint: apply the diffraction equation $n\lambda = d \sin\theta$.)